Elements of Linear Spaces

by

A. R. Amir-Moéz

A. L. Fass

Second Printing 1961

Lithoprinted in U.S.A.
EDWARDS BROTHERS, INC.
Ann Arbor, Michigan

PREFACE

There seems to be a need for an elementary undergraduate course to bridge the gap between freshman mathematics and modern abstract algebra. There are books which attempt an elementary treatment of vector spaces, linear transformations, matrices, etc., but they usually avoid an approach through geometry. Many intelligent students of mathematics feel that their graduate study represents a change of major field, as the methods of generalization are completely new to them.

In this book, we present the elementary material on linear spaces, restricting ourselves to real spaces of dimension not more than three in the first five chapters. Thereafter, the ideas are generalized to complex n-dimensional spaces, and, in chapter 8, we study the geometry of conic sections and quadric surfaces using the methods developed.

A ninth chapter covers the application of the same techniques to problems drawn from various branches of geometry. Following this, two chapters deal with the subject of algebraic structures, especially vector spaces and transformations, from the abstract point of view, including projections and the decomposition of Hermitian operators. A final chapter then treats some of the more accessible recent results on singular values and their use for estimating proper values of general transformations.

We believe both students and instructors are intelligent, and would like to supply details of proof or technique in many places, and have kept this idea in mind in writing the book. We feel also that the book can be used as a quick review for more advanced students.

Problems marked with an asterisk are of greater difficulty or generality, and occasionally require additional background.

The authors would appreciate suggestions and criticism of this text, which may be useful in future revisions.

A. R. A-M., A. L. F.

CONTENTS

Page

PART I

1. REAL EUCLIDEAN SPACE

1.1 **Scalars and vectors:** We call any real number a *scalar.* A line segment AB with an orientation (sense or direction) from **A** to **B** is called the *vector* AB. **A** is called the beginning point and **B** is the end point of AB. By $|AB|$, the *norm* of AB, we mean the length of the line segment **AB**. In this book we consider only vectors having a fixed beginning **O**, called the origin. By a vector V we mean OV. In the special case where **V** is the origin we define the vector V to be the zero vector O. The zero vector may be assumed to have any direction.

1.2 **Sums and scalar multiples of vectors:** Let A and B be two vectors. Then $A + B$ is defined to be the vector R, where **R** is the fourth vertex of the parallelogram whose other vertices are **O**, **A**, and **B** (Fig. 1). In case the two vectors are collinear, the parallelogram degenerates to a line segment (Fig. 2). To add three vectors, we define $A + B + C$ to be R where **R** is that vertex of the parallelepiped determined by the vectors A, B, and C which is not on any edge containing **A**, **B**, or **C** (Fig. 3). The degenerate cases may be considered as before. We observe without proof that addition of three vectors is equivalent to adding one of them to the sum of the other two. Thus we may extend the definition to the sum of any number of vectors, and we observe that this sum is independent of the order of addition, that is, the addition is commutative and associative. By the difference $A - B$ we mean the vector C such that $B + C = A$.

If n is a scalar and A a vector, by nA we mean a vector B such that B is on the line **OA**, if n is positive, **B** and **A** are on the same side of **O**, if n is negative, they are on opposite sides of **O**, and when $n = 0$, $B = O$, and finally, $|B| = |n|\ |A|$. For any two vectors A and B on the same line there is a unique scalar x such that $A = xB$, if $B \neq O$. Comparing this with the degenerate cases of addition, we observe that $(x + y) A = xA + yB$, and $x(yA) = (xy) A$.

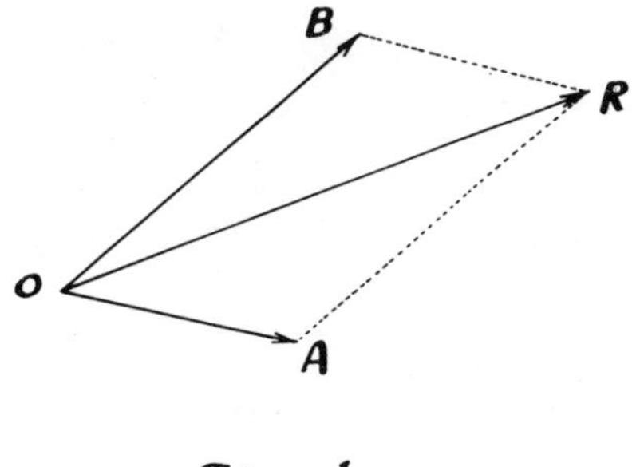

Fig. 1

Fig. 2

1.3 **Linear independence:** Two vectors U and V are called *linearly independent* if the points **O**, **U**, and **V** are not collinear, otherwise they are called *linearly dependent*.

Three vectors U, V, and W are called linearly independent if the points **O**, **U**, **V**, and **W** are not coplanar, otherwise they are linearly dependent. Any four or more vectors are linearly dependent.

1.4 **Theorem:** Let U and V be linearly independent. Given any vector A in the plane **UOV**, two scalars x and y are uniquely defined such that $A = xU + yV$. Conversely given the scalars x and y there is a unique vector A in the plane **UOV** such that $A = xU + yV$.

Proof: The lines through **A**, parallel to U and V respectively, intersect the lines **OV** and **OU** at **C** and **B** (Fig. 4). By 1.2 we have x and y such that $B = xU$ and $C = yV$, and $A = B + C$. Conversely given x and y we get B and C such that $B = xU$ and $C = yV$ and construct A to be $B + C$.

1.5 **Theorem:** Let U, V, and W be linearly independent. Then

(1) none of U, V, and W is in the plane of the other two,
(2) given any vector A, scalars x, y, and z are uniquely defined so that $A = xU + yV + zW$,
(3) given three scalars x, y, and z, there is a unique vector A such that $A = xU + yV + zW$.

Proof: The statement (1) follows trivially from the definition 1.3. The geometric construction of a unique parallelepiped with three sides on the lines **OU**, **OV**, and **OW** with the directions U, V, and W and the diagonal **OA**, and 1.2, make the proof of (2) clear.

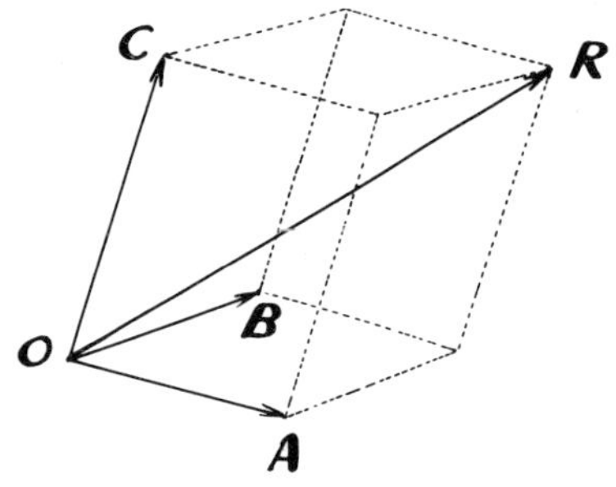

Fig. 3

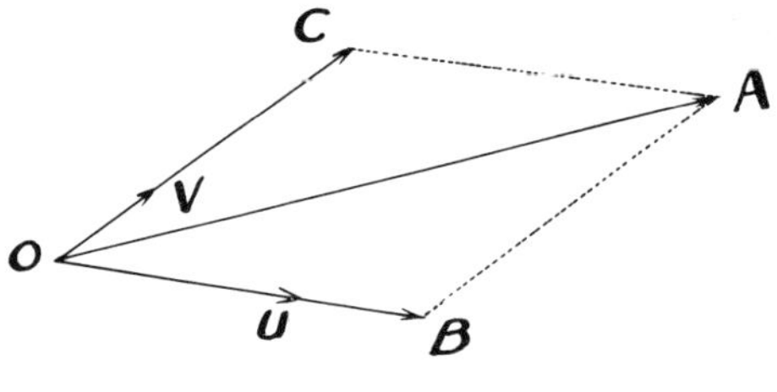

Fig. 4

By 1.2 there are points **B**, **C**, and **D** such that $B = xU$, $C = yV$, $D = zW$, and adding $B + C + D$ gives a unique sum A. This proves (3) (Fig. 5).

Given any set of n vectors $\{U_1, U_2, \ldots, U_n\}$, a vector V is called a *linear combination* of $U_1, U_2, \ldots, U_n$ if there are scalars $a_1, a_2, \ldots, a_n$ such that

$$V = a_1U_1 + a_2U_2 + \ldots + a_nU_n \, .$$

1.6 **Theorem:** A set of vectors is linearly dependent if and only if some one of them can be written as a linear combination of the others.

Proof: For the case of two vectors, if A and B are linearly dependent, then they are collinear, and either $A = xB$ or $B = 0.A$. Also if $A = xB$, then A and B are collinear [see 1.2]. For three vectors A, B, and C, if two are linearly dependent, the proof is as before. If no two are linearly dependent but the three are linearly dependent, then since C is in the plane of A and B, $C = xA + yB$. Also if $C = xA + yB$, then C is in the plane of A and B [see 1.4]. For the case of four or more vectors, again if there are three linearly independent vectors in the set, any other is a linear combination of those three [see 1.5]. Finally any four vectors are linearly dependent [see 1.3].

1.7 **Base (Co-ordinate system):** On any line l let a point **O** as the origin and a vector U, with $|U| \neq 0$, be chosen. Clearly by 1.2, to each vector A on the line l corresponds a scalar x such that $A = xU$ and to each scalar x corresponds a vector A on the line l such that $A = xU$. We call $\{U\}$ a *base* for the line l (Fig. 6), and x the corresponding *component* of the vector A.

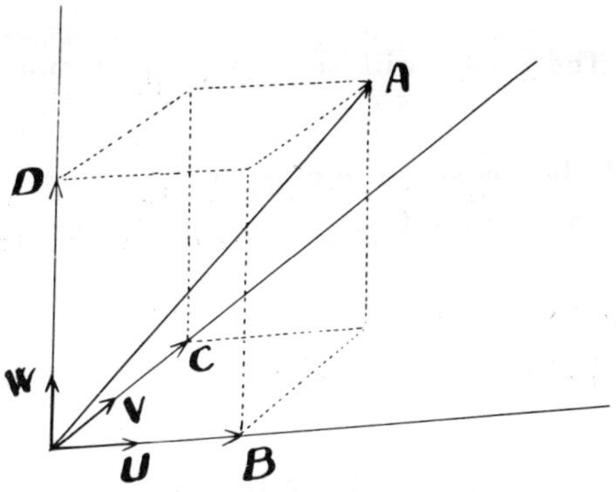

Fig 5

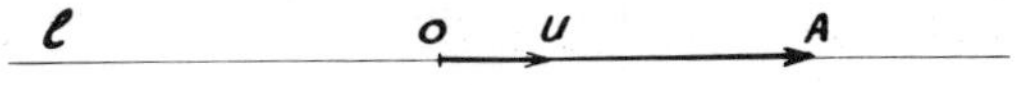

Fig. 6

For a plane, consider two vectors U and V which are linearly independent. By 1.4 for any A there are scalars x and y such that $A = xU + yV$, and given any x and y there is an A such that $A = xU + yV$. We call the set $\{U,V\}$ a *base* for the plane **OUV**, and we order the set, calling U the *first* and V the *second* element of the base. In this way A corresponds to a unique *ordered pair* (x,y) and (x,y) corresponds to a unique vector A (Fig. 4). Again the scalars x and y will be called the *components* of A with respect to the base $\{U,V\}$.

For the whole space we choose U, V, and W linearly independent. By 1.4 for any A there are unique scalars x, y, and z such that $A = xU + yV + zW$ and, conversely, given x, y, and z there is a unique A such that $A = xU+yV+zW$. Exactly as above for the case of the plane, we can establish a correspondence between

A and the *ordered triple* (x,y,z) which is called a *one-to-one* correspondence. The set $\{U,V,W\}$ is called a *base* for the space (Fig. 5), and the scalars x,y, and z the *components* of A with respect to that base.

The choice of vectors for a base on a line or a plane or the space is arbitrary except for the number of elements. Because of this we can introduce the idea of *dimension.* A line is one-dimensional, a plane is of two dimensions, and the space has three dimensions, corresponding to the number of elements in the base. We may represent a vector by the ordered set of its components with respect to a base. These numbers are also called *coordinates* of the end point of the vector. We may therefore denote a vector by the set of its components with respect to a base.

If $\{U,V,W\}$ is a base for the space, we observe that

$$U = 1.U + 0.V + 0.W,$$

hence the components of U with respect to this base are (1,0,0). We write $U = (1,0,0)$. Similarly,

$$V = 0.U + 1.V + 0.W,$$

$$W = 0.U + 0.V + 1.W,$$

hence $V = (0,1,0)$ and $W = (0,0,1)$. The reader will see readily that in a plane with base $\{U,V\}$, $U = (1,0)$ and $V = (0,1)$.

1.8 **Theorem:** Let a base $\{U,V,W\}$ be chosen in the space. Let the vectors $A = x_1U + y_1V + z_1W$ and $B = x_2U + y_2V + z_2W$ be denoted by (x_1,y_1,z_1) and (x_2,y_2,z_2) respectively. Then

$$\text{I} \quad (x_1,y_1,z_1) + (x_2,y_2,z_2) = (x_1 + x_2,\ y_1 + y_2,\ z_1 + z_2) \ ,$$

$$\text{II} \quad a(x_1,y_1,z_1) = (ax_1,ay_1,az_1) \ , \quad a \text{ is a scalar.}$$

Proof: I.

$$(x_1,y_1,z_1) + (x_2,y_2,z_2) = x_1U + y_1V + z_1W + x_2U + y_2V + z_2W =$$

$$(x_1 + x_2)\,U + (y_1 + y_2)\,V + (z_1 + z_2)\,W = (x_1 + x_2,\ y_1 + y_2,\ z_1 + z_2) \ .$$

Part II. follows similarly [see 1.2]. The reader may supply a geometric proof.

Illustration 1: With respect to some base, let $A = (1,-1,2)$, and $B = (2,0,-1)$. Find $3A - 2B$.

By 1.8, $3A - 2B = 3(1,-1,2) - 2(2,0,-1) = (3,-3,6) + (-4,0,2) = (-1,-3,8)$.

Illustration 2: Find the coordinates of the midpoint of the line segment joining the end points of the vectors $A = (x_1,y_1,z_1)$ and $B = (x_2,y_2,z_2)$.

Since the diagonals of a parallelogram bisect each other, the midpoint of the line segment is also the midpoint of the other diagonal $R = A + B$ [see Fig. 1]. Thus the vector $\frac{1}{2}R$ ends at the desired point. Since

$$\frac{1}{2}R = \frac{1}{2}(A + B) = \left(\frac{x_1 + x_2}{2}, \frac{y_1 + y_2}{2}, \frac{z_1 + z_2}{2}\right) ,$$

these numbers are the coordinates of the midpoint.

Illustration 3: Determine whether the vectors (2,3,0), (1,0,2), and (6,-1,0) are linearly dependent or independent.

By 1.6, we observe that (2,3,0) and (1,0,2) are linearly independent since there is no scalar x for which

$$(2,3,0) = x(1,0,2), \text{ that is, } \quad 2 = x, \quad 3 = 0(x), \quad 0 = 2x.$$

We wish to determine, therefore, if there exist scalars x and y such that

$$(6,-1,0) = x(2,3,0) + y(1,0,2) \ ,$$

that is, such that

$$\begin{cases} 2x + y = 6 \\ 3x + 0y = -1 \\ 0x + 2y = 0 \ . \end{cases}$$

A solution of two of these equations does not satisfy the third, hence no such scalars exist. The three vectors are therefore linearly independent.

Illustration 4: Let U, V, and W be the vectors in the space ending at the points with coordinates (2,3,0), (1,0,2), and (6,-1,0). Find the components of the vector A ending at (-1,7,2) with respect to the base $\{U, V, W\}$.

We have already shown (Illustration 3) that the vectors U, V, and W are linearly independent. Thus we must find scalars x, y, z such that

$$A = xU + yV + zW, \qquad \text{that is,}$$

$$\begin{cases} 2x + y + 6z = -1 \\ 3x \qquad - z = 7 \\ \quad 2y \qquad = 2 \ . \end{cases}$$

This system of equations gives the solution $x = 2$, $y = 1$, $z = -1$, hence

$$A = 2U + V - W \ .$$

1.9 **Inner product of two vectors:** Let U and V be two vectors. Let the angle between U and V be α. Then the *inner product* of U and V, denoted by (U, V), is defined to be $|U| \cdot |V| \cos \alpha$. It is clear that

$$(U, V) = (V, U) \quad \text{and} \quad |U|^2 = (U, U),$$

and if U and V are perpendicular

$$(U, V) = 0 \ .$$

We use the word *orthogonal* as synonymous with perpendicular.

1.10 **Projection of a vector on an axis:** By an *axis* we mean a straight line through the origin **O** with a direction, as in analytic geometry. The projection of V on the axis **Ox** is the algebraic (signed) length of **OA**, where **A** is the foot of the perpendicular through **V** to **Ox** (Fig. 7). If V makes an angle α with the

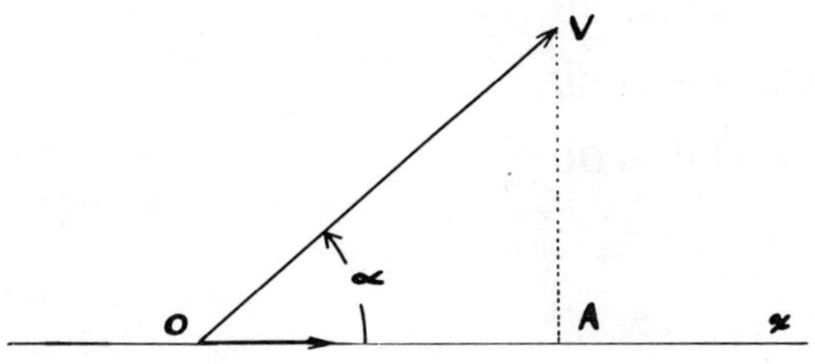

Fig. 7

positive direction of **Ox**, clearly

$$\text{projection of } V \text{ on } \mathbf{Ox} = |V| \cos \alpha .$$

1.11 **Theorem:** Let **Ox** be an axis and U and V two vectors, and let $R = U + V$. Then the projection of R on **Ox** is the sum of the projections of U and of V.

Proof: By constructing three planes through U, V, and R perpendicular to **Ox** we get A, B, and C, the projections of the points U, V, and R respectively (Fig. 8). Comparing the three projections, the result is clear.

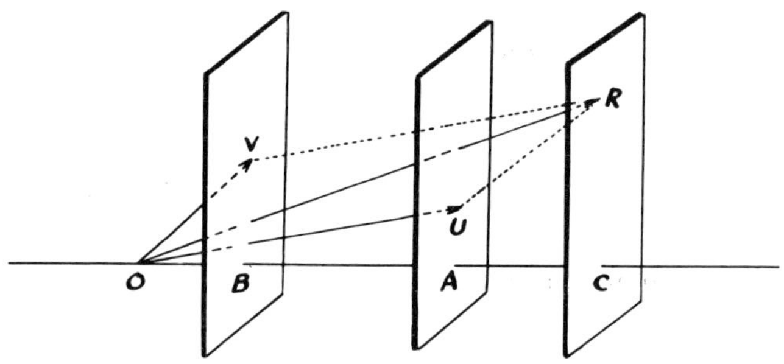

Fig. 8

1.12 **Theorem:** The inner product is distributive with respect to vector addition, i.e., given U, V, and W we have $(U, [V + W]) = (U,V) + (U,W)$.

Proof: Let the angle between U and V be α, the angle between U and W be β, and the angle between $R = V + W$ and U be γ (Fig. 9). Then by 1.10 we have

$$|V| \cos \alpha = \text{projection of } V \text{ on } \mathbf{OU},$$

$$|W| \cos \beta = \text{projection of } W \text{ on } \mathbf{OU},$$

$$|R| \cos \gamma = \text{projection of } R \text{ on } \mathbf{OU}.$$

By 1.11 we have

$$|R| \cos \gamma = |V| \cos \alpha + |W| \cos \beta.$$

Multiplying this by $|U|$ we get

$$|U|\ |V| \cos \alpha + |U|\ |W| \cos \beta = |U|\ |R| \cos \gamma\, \alpha , \quad \text{or}$$

$$(U,R) = (U,[V + W]) = (U,V) + (U,W) .$$

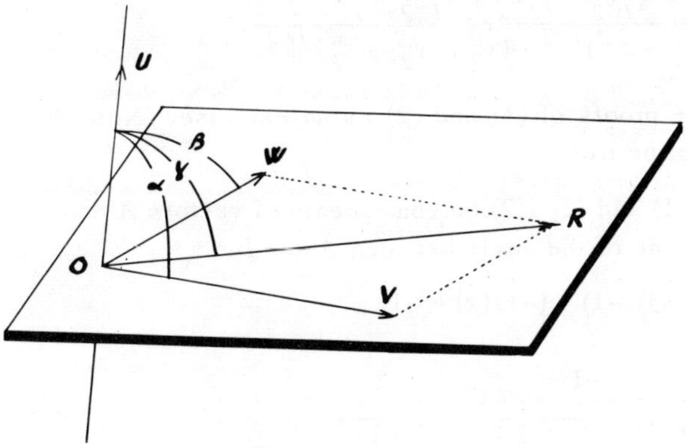

Fig. 9

1.13 **Theorem:** Let U and V be two vectors and c be a scalar. Then

$$(cU,V) = (U,cV) = c(U,V) \quad .$$

Proof: The two cases $c > 0$ and $c < 0$ should be considered separately. The proof is left to the reader as an exercise.

1.14 **Orthonormal base:** If the elements of a base are mutually perpendicular, and the norm of each element is one, then we call the base *orthonormal*. That is, if $\{U, V, W\}$ is orthonormal, then

$$(U,U) = (V,V) = (W,W) = 1 \quad , \quad \text{and}$$

$$(U,V) = (U,W) = (V,W) = 0 \quad .$$

We note that this corresponds to the familiar standard choice of a coordinate system in analytic geometry. To each vector A corresponds an ordered triple (x,y,z) which is the set of ordinary coordinates of the point **A**. We call them also components of A. Here we see that

$$x = (A,U),\ y = (A,V), \text{ and } z = (A,W).$$

Therefore

$$A = (A,U)\,U + (A,V)\,V + (A,W)\,W \quad .$$

1.15 **Norm of a vector and angle between two vectors in terms of components:** Let A be a vector. Clearly $|A| = (A,A)^{1/2}$. If α is the angle between A and B, then by 1.9,

$$\cos\alpha = \frac{(A,B)}{|A| \cdot |B|} \quad .$$

Let $A = x_1U + y_1V + z_1W$ and $B = x_2U + y_2V + z_2W$, where $\{U, V, W\}$ is an orthonormal base in the space. Then

$$|A| = [(x_1U + y_1V + z_1W,\ x_1U + y_1V + z_1W)]^{1/2} = [x_1^2 + y_1^2 + z_1^2]^{1/2} \quad .$$

This is proved by distributivity of inner product with respect to vector addition [see 1.12], and the fact that the base $\{U, V, W\}$ is orthonormal. Similarly it can be proved that

(1) $(A,B) = x_1x_2 + y_1y_2 + z_1z_2$, and

(2) $$\cos \alpha = \frac{x_1x_2 + y_1y_2 + z_1z_2}{(x_1^2 + y_1^2 + z_1^2)^{1/2} \cdot (x_2^2 + y_2^2 + z_2^2)^{1/2}}.$$

The reader should supply the proofs of (1) and (2) as an exercise. Note that if the base is not orthonormal what was said in 1.15 will not be true.

Illustration 1: Let (2,3,-1) and (2,-1,2) be components of vectors A and B with respect to an orthonormal base. Find (A,B) and the cosine of the angle between A and B.

$$(A,B) = (2)(2) + (3)(-1) + (-1)(2) = -1 \ ,$$

$$\cos = -\frac{-1}{\sqrt{4+9+1} \cdot \sqrt{4+1+4}} = \frac{-1}{3\sqrt{14}}.$$

Illustration 2: Find a vector which is perpendicular to the vectors in illustration 1, and has norm $5\sqrt{5}$.

Let the desired vector be $(x, y, z) = C$. Then

$$(A,C) = (B,C) = 0, \qquad (C,C) = 125. \text{ Thus}$$

$$\begin{cases} 2x + 3y - z = 0 \\ 2x - y + 2z = 0 \\ x^2 + y^2 + z^2 = 125 \ . \end{cases}$$

Solving this system, we find the two solutions

(5,-6,-8) and (-5,6,8,) .

Illustration 3: Find the distance between the points $\mathbf{A} = (x_1,y_1,z_1)$ and $\mathbf{B} = (x_2,y_2,z_2)$. The distance is equal to the norm of the vector $B - A$ [see 1.2].

$$|B - A|^2 = (B - A, B - A).$$

But

$$B - A = (x_2 - x_1, y_2 - y_1, z_2 - z_1).$$

hence by 1.15,

$$|B - A| = [(x_2 - x_1)^2 + (y_2 - y_1)^2 + (z_2 - z_1)^2]^{1/2} .$$

1.16 **Orthonormalization of a base:** We can construct an orthonormal base out of any base $\{U, V, W\}$ as follows:

Take $$U_1 = \frac{1}{|U|} U \ ,$$

$$V_1 = \frac{1}{|V - (V, U_1) U_1|} [V - (V, U_1) U_1], \text{ (Fig. 10), and}$$

$$W_1 = \frac{1}{|W - [(W, U_1) U_1 + (W, V_1) V_1]|} \{W - [(W, U_1) U_1 + (W, V_1) V_1]\}, \text{ (Fig. 11).}$$

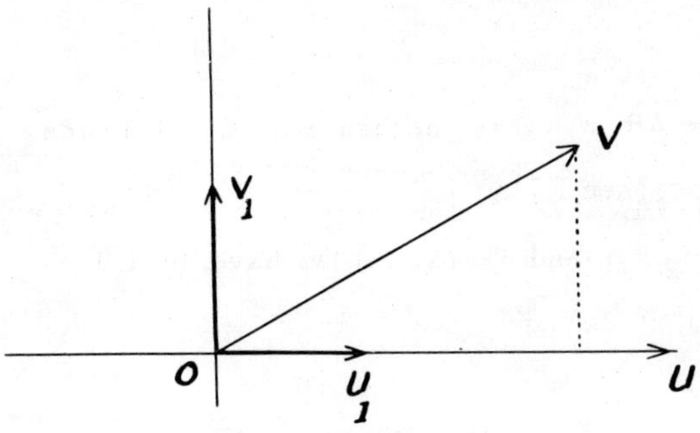

Fig. 10

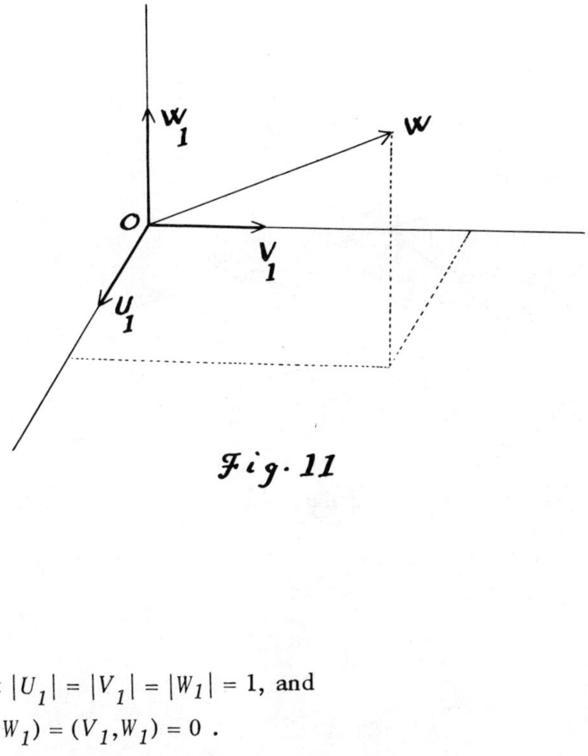

Fig. 11

The reader should show that $|U_1| = |V_1| = |W_1| = 1$, and

$$(U_1,V_1) = (U_1,W_1) = (V_1,W_1) = 0 \ .$$

1.17 **Subspaces:** A line through the origin is called a *one-dimensional linear subspace* of the whole space or of a plane containing it. A plane through the origin is called a *two-dimensional subspace* of the whole space. Such a line contains all scalar multiples of any non-zero vector whose end is on the line. A plane through the origin consists of all linear combination of any two linearly independent vectors in the plane.

1.18 **Straight line:** Given two points **A** and **B** in the space, we consider the vector $V = B - A$, (Fig. 12), and we call the components l, m, and n of V with respect to a given base the *direction numbers* of the straight line through **A** and **B**. The set (l,m,n) is an ordered set, and we call l, m, and n respectively the x, y, and z direction numbers of the line through **A** and **B**. If V is of unit length, and the base is orthonormal, we call the components of V *direction cosines* of the line through **A** and **B**, and we denote them by λ, μ, and ν. Clearly

$$\lambda^2 + \mu^2 + \nu^2 = 1 .$$

For any other point **W** on the line **AB**, $W - A$ is collinear with $B - A$, hence

$$W - A = t(B - A) \qquad \text{[see 1.2].}$$

Thus if $A = (x_1,y_1,z_1)$, $B = (x_2,y_2,z_2)$, and $W = (x,y,z)$, we have, by 1.8,

$$(1) \qquad \begin{cases} x - x_1 = t(x_2 - x_1) \\ y - y_1 = t(y_2 - y_1) \\ z - z_1 = t(z_2 - z_1) \end{cases}$$

as a set of parametric equations of the line **AB**. We can also suppose that

$$l = x_2 - x_1, \qquad m = y_2 - y_1, \qquad \text{and} \qquad n = z_2 - z_1,$$

and we write (1) as

$$(2) \qquad \begin{cases} x = x_1 + tl \\ y = y_1 + tm \\ z = z_1 + tn . \end{cases}$$

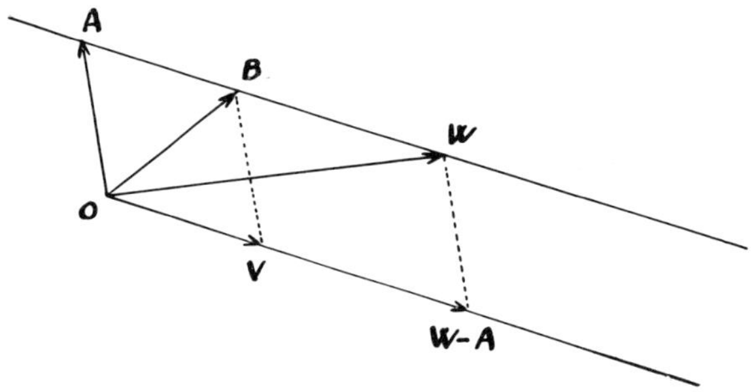

Fig. 12

In vector form, we have

$$W = A + t(B - A) .$$

If l, m, and n are all different from zero we can write (2) as

$$t = \frac{x - x_1}{l} = \frac{y - y_1}{m} = \frac{z - z_1}{n} .$$

It is clear that the form of the equations of a line is independent of the orthonormality of the base.

1.19 **Plane:** Let a base for the space be chosen orthonormal. Let

$$V = (l,m,n)$$

be a vector perpendicular to the plane $\mathcal{P}$. Let $A = (x_o,y_o,z_o)$ end on $\mathcal{P}$. For any point P of the plane, let $P = (x,y,z)$. Then the equation of this plane may be obtained by writing that the vector (l,m,n) is perpendicular to $P - A$ (Fig. 13). That is,

(1) $$l(x - x_o) + m(y - y_o) + n(z - z_o) = 0 \qquad \text{[see 1.9, 1.15]},$$

or $$lx + my + nz + d = 0\,, \quad \text{where} \quad d = -lx_o - my_o - nz_o\,.$$

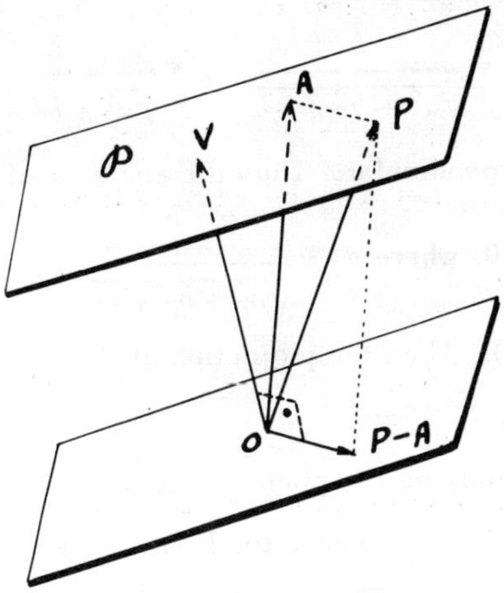

Fig. 13

In vector form, (1) may be written as

(2) $$(V, P - A) = 0\,.$$

Illustration: Find the equation of the plane which contains the line

$$\frac{x-1}{1} = \frac{y+2}{2} = \frac{z-1}{3}$$

and is perpendicular to the plane $x - y + 2z + 1 = 0$.

The desired plane must contain a line perpendicular to the plane $x - y + 2z + 1 = 0$. That is, a line with direction numbers (1,-1,2). Thus a vector perpendicular to the desired plane must be perpendicular to (1,-1,2) and (1,2,3) which is a vector parallel to the given line. Suppose (l,m,n) is a vector perpendicular to the desired plane. Then by 1.9 and 1.15 we have

$$\begin{cases} l - m + 2n = 0 \\ l + 2m + 3n = 0\,. \end{cases}$$

Since we are looking for one vector we can solve this system of equations for l and m in terms of n

$$\begin{cases} l - m = -2n \\ l + 2m = -3n\,. \end{cases}$$

This gives

$$3m = -n, \quad \text{or} \quad m = -\frac{n}{3} \quad \text{and} \quad 3l = -7n, \quad \text{or} \quad l = -\frac{7n}{3}.$$

If we choose $n = -3$, we get (7,1,-3) perpendicular to the desired plane. Since the plane contains the line

$$x - 1 = \frac{y+2}{2} = \frac{z-1}{3}$$

it passes through (1,-2,1). Thus the equation of the plane is

$$7(x-1)+(y+2)-3(z-1)=0\ .$$

1.20 **Distance between a point and a plane:** Let **A** be a point such that $A=(x_o,y_o,z_o)$, and let $ax+by+d=0$ be a plane. Then the distance of **A** from the plane is

$$\delta=\left|\frac{ax_o+by_o+cz_o+d}{\sqrt{a^2+b^2+c^2}}\right|.$$

Proof: Clearly a, b, c are direction numbers of a line perpendicular to the plane. Then

$$\lambda=\frac{a}{\sqrt{a^2+b^2+c^2}},\quad \mu=\frac{b}{\sqrt{a^2+b^2+c^2}},\quad \nu=\frac{c}{\sqrt{a^2+b^2+c^2}}$$

are the direction cosines of this perpendicular. Thus the equation of the plane can be written as

$$\lambda\, x+\mu\, y+\nu\, z+d_1=0,\ \text{where } d_1=\frac{d}{\sqrt{a^2+b^2+c^2}}.$$

Let **Op** be an axis containing (λ,μ,ν). Then the projection of A on **Op** is

$$\delta_1=\lambda x_o+\mu y_o+\nu z_o\ ,$$

and the projection of B, a vector ending on the plane, on **Op** is

$$\delta_2=\lambda x+\mu y+\nu z=-d_1 \qquad [\text{see } 1.10,\ 1.15]\ .$$

Therefore

$$\delta=|\delta_2-\delta_1|=|\lambda x_o+\mu y_o+\nu z_o+d_1|=\left|\frac{ax_o+by_o+cz_o+d}{\sqrt{a^2+b^2+c^2}}\right|$$

Illustration: Find the equation of a tangent plane to the sphere

$$(x-1)^2+(y-2)^2+(z-3)^2=4$$

which is parallel to the plane $3x-6y+2z=0$.

The desired plane is of the form

$$3x-6y+2z+p=0.$$

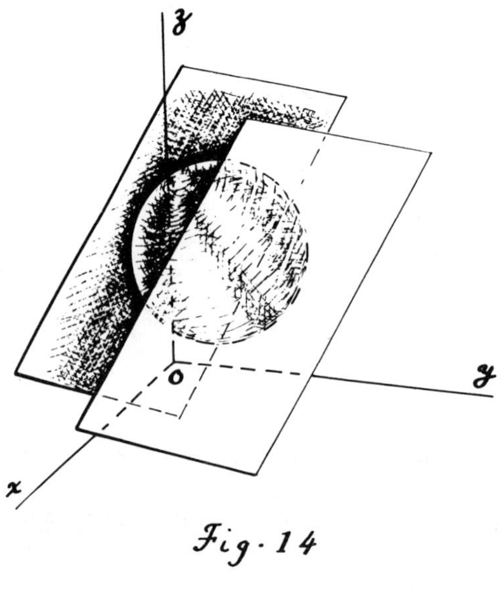

Fig. 14

We have to find p such that the distance of the center of the sphere from this plane will be equal to the radius of the sphere, which is 2. (Fig. 14). Thus

$$\left|\frac{3 - 12 + 6 + p}{7}\right| = 2 .$$

This implies that

$$\frac{p - 3}{7} = \pm 2 .$$

Therefore either $p = 17$ or $p = -11$. The tangent planes are

$$3x - 6y + 2z + 17 = 0 \quad \text{and} \quad 3x - 6y + 2z - 11 = 0 .$$

EXERCISES 1

1. Prove that for three vectors A, B, and C, $A + B = B + A$, and $(A + B) + C = A + (B + C)$.
2.* Prove that the sum of any finite number of vectors is independent of the order of adding.
3. Examine each set of vectors for linear independence:

   (i) (1,1,0), (0,1,1), (3,0,0) ;
   (ii) (2,0), (0,5) ;
   (iii) (5,2,9), (2,-1,-1), (7,1,8) ;
   (iv) (0,0,1), (2,0,0), (-2,0,1) .

4. Show that (a_1, a_2) and (b_1, b_2) are linearly independent if and only if $a_1 b_2 - a_2 b_1 = 0$.
5. If V_1, V_2, V_3 are the vectors of 3(i), find $V_1 - V_2$ and $3V_1 + 2V_2 - V_3$.

   In the following problems, assume all bases are orthonormal.

6. Find the angle between:

   (i) (3,-1), (5,-2) ;
   (ii) (-1,-1,0), (5,0,-2) ;
   (iii) (2,-7,7), (0,0,5) .

7. Find a vector of norm one perpendicular to (-4,3).
8. Given $U = (0,2,2)$ and $V = (1,2,0)$ find a vector W orthogonal to U and V.
9. Orthonormalize:

   (i) (2,0), (1,1) ;
   (ii) (0,1,-1), (5,6,0), (2,0,0) ;
   (iii) (2,-1,-3), (-1,5,1) .

10. Show that {(5,2,0), (2,1,0)} is a base for a subspace. Find the equation of the subspace. Orthonormalize this base.
11. Show that (1,1,0), (0,0,2), and (0,3,2) are linearly independent. Find the components of (5,9,-2) with respect to the base consisting of those vectors.
12. If $U_1, U_2, \ldots, U_k$ are any vectors, show that the set of all vectors of the form $a_1 U_1 + a_2 U_2 + \ldots + a_k U_k$ forms a subspace or the whole space.
13.* The equation of a surface with respect to a rectangular coordinate system is $f(x,y,z) = 0$.

   (i) Find the components of a vector in the tangent plane at the point (x,y,z) of the surface with respect to an orthonormal base whose origin is at (x,y,z) and for which U, V, and W of the base are parallel to the coordinate axes.
   (ii) Find the components of a vector normal to the surface at (x,y,z) with respect to the same base. (Hint: $df = 0$)

14.* Show that the angle α between two plane curves $f(x,y) = 0$ and $g(x,y) = 0$ at (x,y), the point of intersection is:

$$\alpha = \text{Arc cos}\, \frac{\dfrac{\partial f}{\partial x}\dfrac{\partial g}{\partial x} + \dfrac{\partial f}{\partial y}\dfrac{\partial g}{\partial y}}{\sqrt{\left(\dfrac{\partial f}{\partial x}\right)^2 + \left(\dfrac{\partial f}{\partial y}\right)^2} \cdot \sqrt{\left(\dfrac{\partial g}{\partial x}\right)^2 + \left(\dfrac{\partial g}{\partial y}\right)^2}}.$$

15.* Generalize 14 for the angle between two surfaces in the space.

16. Show that the normal equation of a line in the plane can be found by the use of inner product of a vector whose end is on the line and a vector of unit norm on the perpendicular from the origin to the line.

17. Generalize 16 for a plane in the space.

In the following problems whenever necessary, choose the base to be orthonormal.

18. Find the equation of the plane:

(i) through (3,1,-2) perpendicular to a line with direction numbers (2,-2,3) ;
(ii) through (3,1,-2) parallel to $2x + 5y - z = 2$;
(iii) through (1,-1,2), (3,2,1), (-1,-3,3) ;
(iv) whose intercepts are 3, 1, -4;
(v) which is the perpendicular bisector of the line segment joining (1,-3,-5) and (3,1,1) ;
(vi) which has equal intercepts and passes through the point (3,7,-2).

19. Find equations of the straight line

(i) passing through (5,1,8) and (7,4,5) ;
(ii) passing through (1,-3,-5) and (3,1,1) ;
(iii) passing through (-2,1,0), perpendicular to the plane $3x - 2y + z = 1$;
(iv) passing through (1,-3,-5) and parallel to

$$\frac{x-2}{2} = \frac{y+1}{3} = \frac{z}{4} ;$$

(v) passing through (2,1,-1) and parallel to the planes

$$2x + 3y - z = 6 \quad \text{and} \quad x + y + z = 3 .$$

20. Find the distance between (3,7,-9) and $x - y - z - 1 = 0$.

21. Find the distance between the pair of parallel planes

$$x - 2y - 2z + 1 = 0 \quad \text{and} \quad x - 2y - 2z + 10 = 0.$$

22. Find the projection of

(i) the point (1,0,0) on the line $x = y = z$;
(ii) the point (7,4,-5) on the line

$$\begin{cases} x = -13 + 5t \\ y = 8 - 3t \\ z = -6 + 2t . \end{cases}$$

23. Write the equation of a plane tangent to the sphere $x^2 + y^2 + z^2 = 9$ which contains the line

$$\begin{cases} x = 5 \\ z = 0 . \end{cases}$$

24. Find the area of the triangle whose vertices are (2,-1,3), (4,3,-2), and (3,0,-1) (base is orthonormal).
25. Find parametric equations of the line of intersection of the planes

$$3x - y + 3z - 2 = 0 \quad \text{and} \quad x + 2y - 3z + 4 = 0 .$$

ADDITIONAL PROBLEMS

1. Show that $(cA,B) = c(A,B)$, where c is a scalar.
2. Supply the details in 1.9.
3. Let (l,m) be the set of direction numbers of a straight line with respect to an orthonormal base. Show that the slope of the line is m/l.
4. Find conditions on (x,y,z) such that (x,y,z) is a vector of length 5 perpendicular to $V = (1,7,-2)$. (Base is orthonormal).
5. Find a vector of length 3 orthogonal to (1,2,7) and (-2,-4,-14). Explain the result.
6. Find the components of (5,2) with respect to the base {(1,-1), (3,5)}.
7. Find the components of (1,-1,7) with respect to the base {(1,1,2), (3,-1,0), (2,0,-11)}.
8. Find the equation of the subspace:

   (i) containing (2,1,0) and (7,0,1) ;
   (ii) of one dimension and containing (1,1,9) ;
   (iii) containing (1,-1,2) and perpendicular to $x = y$.

9. Show that {(6/7,-3/7,2/7), (2/7,6/7,3/7), (-3/7,-2/7,6/7)} forms an orthonormal base.
10. Find the equation of the sphere with center (4,1,-2) tangent to the plane $2x - y + 2z + 9 = 0$.
11. Find the equation of the plane that contains the line

$$\frac{x-1}{2} = \frac{y+2}{3} = z$$

and is parallel to the line

$$\frac{x+2}{3} = \frac{y-1}{3} = \frac{z-2}{-1} .$$

12. Find the equation of a tangent plane to the sphere

$$(x-2)^2 + (y+1)^2 + (z+2)^2 = 9$$

which is parallel to $x - y + 2z = 0$.

13. Show that in the plane any vector of norm one can be written as $(\cos\alpha, \sin\alpha)$, where α is the angle that the vector makes with the x-axis.
14. Let two vectors of norm one in the plane be $(\cos\alpha, \sin\alpha)$ and $(\cos\beta, \sin\beta)$. Using inner product of these vectors obtain a formula for $\cos(\alpha - \beta)$.
15. Find the area of a triangle whose vertices are

   (i) (3,1,2), (4,5,10), and (11,-3,3) ;
   (ii) (2,5,4), (3,6,8), and (2,10,-1).

16. Let the vector P vary, ending on the line

$$x = \frac{y-2}{3} = \frac{z-1}{2} .$$

Find a condition on (x,y,z) such that $V = (x,y,z)$ will satisfy
   (1) $V - P$ is perpendicular to the given line;
   (2) $|V - P| = 3$

17. Find the equation of the surface formed by revolving the line $x = 0$, $y = 1$ about the line $x = \frac{y}{2} = \frac{z}{3}$.

2. LINEAR TRANSFORMATIONS AND MATRICES

2.1 **Definition:** A *linear transformation* $\mathbf{A}$ on the space is a method of corresponding to each vector V of the space another vector $V_1 = \mathbf{A}\,V$ of the space, such that for any vectors U and V and any scalars a and b,

(1) $\mathbf{A}(aU + bV) = a\mathbf{A}U + b\mathbf{A}V.$

Remark: In what follows all transformations referred to will be linear.

2.2 **Addition and Multiplication of Transformations:** Let $\mathbf{A}$ and $\mathbf{B}$ be two transformations on the space. Then we define the transformation $\mathbf{C} = \mathbf{A} + \mathbf{B}$, by the relation $\mathbf{C}V = \mathbf{A}V + \mathbf{B}V$ for every vector V. The transformation $\mathbf{C}$ is called the *sum* of $\mathbf{A}$ and $\mathbf{B}$. The *product* $\mathbf{AB}$ of two transformations $\mathbf{A}$ and $\mathbf{B}$ is defined by the relation $(\mathbf{AB})V = \mathbf{B}(\mathbf{A}V)$ for every V, i.e., if $\mathbf{A}V = V_1$, then $(\mathbf{AB})V = \mathbf{B}V_1$. If k is a scalar, we define $k\mathbf{A}$ by the relation $(k\mathbf{A})V = k(\mathbf{A}V)$. The reader may verify the fact that $\mathbf{A} + \mathbf{B}$, $\mathbf{AB}$, and $k\mathbf{A}$ are transformations satisfying 2.1,(1).

2.3 **Theorem:** The multiplication of transformations is associative; also for transformations the multiplication is distributive with respect to addition. That is, for any three transformations $\mathbf{A}$, $\mathbf{B}$, and $\mathbf{C}$:

(1) $\mathbf{A}(\mathbf{BC}) = (\mathbf{AB})\mathbf{C},$

(2) $\mathbf{A}(\mathbf{B} + \mathbf{C}) = \mathbf{AB} + \mathbf{AC}.$

Proof: By 2.2, $[\mathbf{A}(\mathbf{BC})]V = (\mathbf{BC})(\mathbf{A}V) = \mathbf{C}[\mathbf{B}(\mathbf{A}V)]$ for any V. On the other hand $[(\mathbf{AB})\mathbf{C}]V = \mathbf{C}[(\mathbf{AB})V] = \mathbf{C}[\mathbf{B}(\mathbf{A}V)]$.

This proves (1).

Also by 2.2, for any V we have

$[\mathbf{A}(\mathbf{B} + \mathbf{C})]V = (\mathbf{B} + \mathbf{C})(\mathbf{A}V) = \mathbf{B}(\mathbf{A}V) + \mathbf{C}(\mathbf{A}V) = (\mathbf{AB})V + (\mathbf{AC})V = (\mathbf{AB} + \mathbf{AC})V.$ Thus

$$\mathbf{A}(\mathbf{B} + \mathbf{C}) = \mathbf{AB} + \mathbf{AC}.$$

2.4 **Matrix of a Transformation A:** Let $\{U_1, U_2, U_3\}$ be a base for the space, not necessarily orthonormal. Suppose for the transformation $\mathbf{A}$,

$$\mathbf{A}U_1 = a_{11}U_1 + a_{12}U_2 + a_{13}U_3\,,$$
$$\mathbf{A}U_2 = a_{21}U_1 + a_{22}U_2 + a_{23}U_3\,,$$
$$\mathbf{A}U_3 = a_{31}U_1 + a_{32}U_2 + a_{33}U_3\,.$$

We see that (a_{11}, a_{12}, a_{13}), (a_{21}, a_{22}, a_{23}), and (a_{31}, a_{32}, a_{33}) are respectively the components of $\mathbf{A}U_1$, $\mathbf{A}U_2$, and $\mathbf{A}U_3$. We put these components respectively in three rows as follows

$$\begin{pmatrix} a_{11} & a_{12} & a_{13} \\ a_{21} & a_{22} & a_{23} \\ a_{31} & a_{32} & a_{33} \end{pmatrix},$$

and call this array of numbers the *matrix* of $\mathbf{A}$ with respect to the base $\{U_1, U_2, U_3\}$, and we denote it by matrix $\mathbf{A}$.

The choice of two subscripts for each scalar indicates to what row and what column of the matrix it belongs; for example a_{23} means the scalar situated in the second row and the third column.

It is clear that $\mathbf{A}$ on a two-dimensional space, i.e., the plane, has a matrix of two rows and two columns, and on a line, a one-by-one matrix.

Illustration 1: Let **A** be the 30° rotation of every vector about the origin in the plane **xOy**. Then

1. **A** is a linear transformation.

Proof: Suppose U and V are two vectors. Let a and b be two scalars and $R = aU + b\,V$ (Fig. 15). Let $P = aU$ and $Q = bV$. It is clear that after the 30° rotation the parallelogram **OPRQ** becomes a congruent parallelogram, i.e.,

$$\mathbf{A}\,(aU + bV) = a\mathbf{A}U + b\mathbf{A}V.$$

2. The matrix of **A** with respect to the base $\{(1,0),\ (0,1)\}$ is found as follows:

$$\mathbf{A}\,(1,0) = \frac{\sqrt{3}}{2}(1,0) + \frac{1}{2}(0,1),$$

$$\mathbf{A}\,(0,1) = -\frac{1}{2}(1,0) + \frac{\sqrt{3}}{2}(0,1), \quad \text{(Fig. 16)}.$$

Therefore the matrix of **A** is

$$\begin{pmatrix} \frac{\sqrt{3}}{2} & \frac{1}{2} \\ -\frac{1}{2} & \frac{\sqrt{3}}{2} \end{pmatrix}.$$

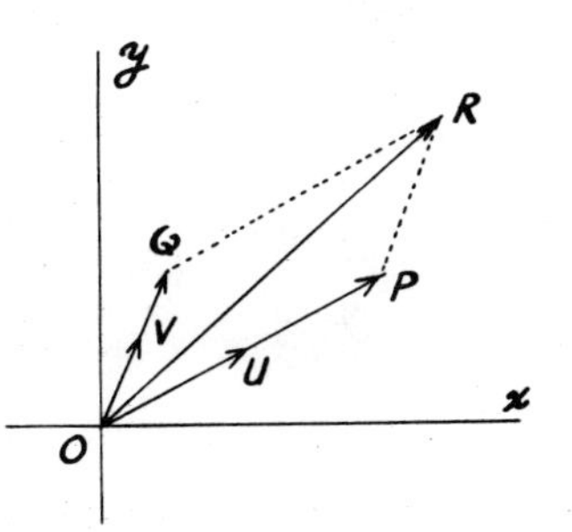

Fig. 15

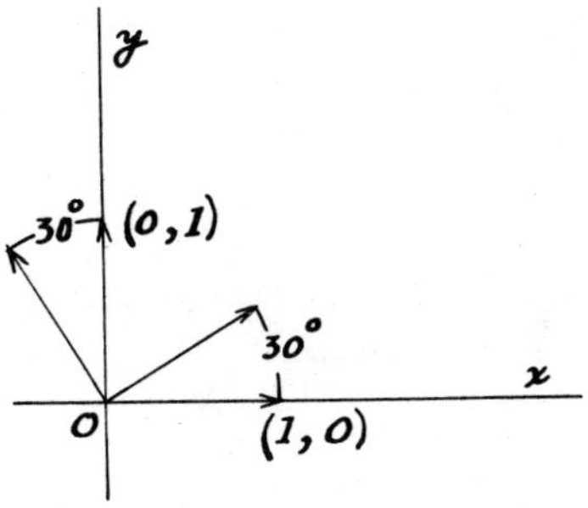

Fig. 16

Illustration 2: For a transformation **A**, we have $\mathbf{A}(x,y) = (x',y')$ such that

$$\begin{cases} x' = 2x \\ y' = x + y\,. \end{cases}$$

Then

I. **A** is linear.

$$\mathbf{A}[a(x_1,y_1) + b(x_2,y_2)] = \mathbf{A}[(ax_1 + bx_2,\ ay_1 + by_2)] = (2ax_1 + 2bx_2,\ ax_1 + ay_1 + bx_2 + by_2) =$$

$$(2ax_1,\ ax_1 + ay_1) + (2bx_2,\ bx_2 + by_2) = a(2x_1,\ x_1 + y_1) + b(2x_2,\ x_2 + y_2) = a\,\mathbf{A}(x_1,y_1) + b\,\mathbf{A}(x_2,y_2)\,.$$

II. The matrix of this transformation with respect to the base $\{(1,0), (0,1)\}$ is found as follows:

$$\mathbf{A}(0,1) = (2,1) = 2(1,0) + 1(0,1)$$

$$\mathbf{A}(0,1) = (0,1) = 0(1,0) + 1(0,1) .$$

Hence the matrix of **A** is

$$\begin{pmatrix} 2 & 1 \\ 0 & 1 \end{pmatrix} .$$

Illustration 3: Let **A** be the perpendicular projection of a vector into the plane $z = 0$. Then

I. **A** is a linear transformation.

Let $R = aV + bW$, $\mathbf{A}R = R_1$, $\mathbf{A}V = V_1$, and $\mathbf{A}W = W_1$. By observation of (Fig. 17) we see that $\mathbf{A}(aV) = aV_1$, $\mathbf{A}(bW) = bW_1$, and $aV_1 + bW_1 = R_1$. A complete geometric proof is simple but lengthy.

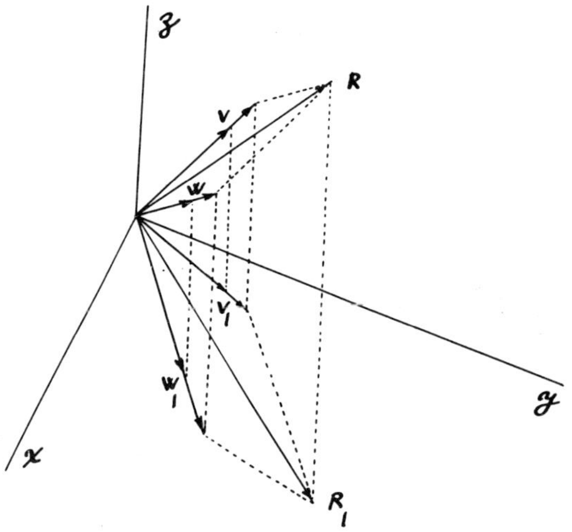

Fig. 17

II. Let the base $\{(1,0,0), (0,1,0), (0,0,1)\}$ be chosen. For the matrix of **A** we observe that $\mathbf{A}(x,y,z) = (x,y,0)$. Thus

$$\mathbf{A}(1,0,0) = (1,0,0)$$

$$\mathbf{A}(0,1,0) = (0,1,0)$$

$$\mathbf{A}(0,0,1) = (0,0,0) ,$$

and the matrix of **A** is

$$\begin{pmatrix} 1 & 0 & 0 \\ 0 & 1 & 0 \\ 0 & 0 & 0 \end{pmatrix}$$

Illustration 4: Let $\mathbf{A}(2,-1) = (5,7)$, and $\mathbf{A}(-3,0) = (0,6)$. Find the matrix of **A** with respect to the base $\{(1,0), (0,1)\}$.

We observe that $(2,-1)$ and $(-3,0)$ are linearly independent. We find scalars a, b, c, and d such that

$$\begin{cases} (1,0) = a(2,-1) + b(-3,0) \\ (0,1) = c(2,-1) + d(-3,0) \,. \end{cases}$$

This gives

$$\begin{cases} 2a - 3b = 1 \\ \quad -a \quad = 0 \\ 2c - 3d = 0 \\ \quad -c \quad = 1 \,, \end{cases}$$

and the solutions are $a = 0$, $b = -1/3$, $c = -1$, and $d = -2/3$. Then

$$\mathbf{A}(1,0) = \mathbf{A}[0(2,-1) - \frac{1}{3}(-3,0)] = 0\,\mathbf{A}(2,-1) - \frac{1}{3}\mathbf{A}(-3,0) = -\frac{1}{3}(0,6) = (0,-2) \,;$$

$$\mathbf{A}(0,1) = \mathbf{A}[-1(2,-1) - \frac{2}{3}(-3,0)] = -1\mathbf{A}(2,-1) - \frac{2}{3}\mathbf{A}(-3,0) = -1(5,7) - \frac{2}{3}(0,6) = (-5,-11) \,.$$

The matrix is therefore

$$\begin{pmatrix} 0 & -2 \\ -5 & -11 \end{pmatrix} .$$

2.5 **Unit and zero transformation:** A transformation is called the *unit* or *identity* transformation, denoted by **I**, when it sends every vector to itself. The matrix of the unit transformation is found as follows:

Let $\{U_1, U_2, U_3\}$ be any base in the space. Then

$$\mathbf{A}U_1 = U_1 = 1(U_1) + 0(U_2) + 0(U_3) \,,$$
$$\mathbf{A}U_2 = U_2 = 0(U_1) + 1(U_2) + 0(U_3) \,,$$
$$\mathbf{A}U_3 = U_3 = 0(U_1) + 0(U_2) + 1(U_3) \,.$$

Therefore

$$\text{matrix } \mathbf{I} = \begin{pmatrix} 1 & 0 & 0 \\ 0 & 1 & 0 \\ 0 & 0 & 1 \end{pmatrix} .$$

The *zero* transformation is the one sending every vector to zero. We denote it by **O**. The matrix of **O** can be found in a similar way, and it is

$$\begin{pmatrix} 0 & 0 & 0 \\ 0 & 0 & 0 \\ 0 & 0 & 0 \end{pmatrix} .$$

2.6 **Addition of Matrices:** Let

$$\begin{pmatrix} a_{11} & a_{12} & a_{13} \\ a_{21} & a_{22} & a_{23} \\ a_{31} & a_{32} & a_{33} \end{pmatrix} \quad \text{and} \quad \begin{pmatrix} b_{11} & b_{12} & b_{13} \\ b_{21} & b_{22} & b_{23} \\ b_{31} & b_{32} & b_{33} \end{pmatrix}$$

be respectively the matrices of the transformations **A** and **B** with respect to the base $\{U_1, U_2, U_3\}$. Let $\mathbf{C} = \mathbf{A} + \mathbf{B}$. Then the matrix of **C** is found as follows:

$$\mathbf{C}U_1 = (\mathbf{A}{+}\mathbf{B})\,U_1 = \mathbf{A}U_1 + \mathbf{B}U_1 = a_{11}U_1 + a_{12}U_2 + a_{13}U_3 + b_{11}U_1 + b_{12}U_2 + b_{13}U_3 \,.$$

Therefore

$$\mathbf{C}U_1 = (a_{11} + b_{11})\,U_1 + (a_{12} + b_{12})\,U_2 + (a_{13} + b_{13})\,U_3. \text{ Similarly}$$
$$\mathbf{C}U_2 = (a_{21} + b_{21})\,U_1 + (a_{22} + b_{22})\,U_2 + (a_{23} + b_{23})\,U_3,$$
$$\mathbf{C}U_3 = (a_{31} + b_{31})\,U_1 + (a_{32} + b_{32})\,U_2 + (a_{33} + b_{33})\,U_3.$$

Thus the matrix of $\mathbf{A} + \mathbf{B}$ is

$$\begin{pmatrix} a_{11} + b_{11} & a_{12} + b_{12} & a_{13} + b_{13} \\ a_{21} + b_{21} & a_{22} + b_{22} & a_{23} + b_{23} \\ a_{31} + b_{31} & a_{32} + b_{32} & a_{33} + b_{33} \end{pmatrix} .$$

We define this matrix to be the *sum* of the matrices of **A** and of **B**.

2.7 **Product of Matrices:** Let

$$\begin{pmatrix} a_{11} & a_{12} & a_{13} \\ a_{21} & a_{22} & a_{23} \\ a_{31} & a_{32} & a_{33} \end{pmatrix} \quad \text{and} \quad \begin{pmatrix} b_{11} & b_{12} & b_{13} \\ b_{21} & b_{22} & b_{23} \\ b_{31} & b_{32} & b_{33} \end{pmatrix} \quad \text{be the}$$

matrices of the transformations **A** and **B** with respect to the base $\{U_1, U_2, U_3\}$. Let $\mathbf{C} = \mathbf{AB}$.

We find the matrix of **C** as follows:

$$\mathbf{A}U_1 = a_{11}U_1 + a_{12}U_2 + a_{13}U_3 \,,$$
$$\mathbf{A}U_2 = a_{21}U_1 + a_{22}U_2 + a_{23}U_3 \,,$$
$$\mathbf{A}U_3 = a_{31}U_1 + a_{32}U_2 + a_{33}U_3 \,.$$

By 2.2 and 2.1 we have

$$\mathbf{B}\,(\mathbf{A}U_1) = a_{11}\mathbf{B}U_1 + a_{12}\mathbf{B}U_2 + a_{13}\mathbf{B}U_3$$
$$= a_{11}(b_{11}U_1 + b_{12}U_2 + b_{13}U_3) + a_{12}(b_{21}U_1 + b_{22}U_2 + b_{23}U_3) + a_{13}(b_{31}U_1 + b_{32}U_2 + b_{33}U_3)$$
$$= (a_{11}b_{11} + a_{12}b_{21} + a_{13}b_{31})\,U_1 + (a_{11}b_{12} + a_{12}b_{22} + a_{13}b_{32})\,U_2 + (a_{11}b_{13} + a_{12}b_{23} + a_{13}b_{33})\,U_3 \,.$$

Similarly $\mathbf{B}(\mathbf{A}U_2)$ and $\mathbf{B}(\mathbf{A}U_3)$ are obtained, and the matrix of **AB** will be:

$$\begin{pmatrix} a_{11}b_{11} + a_{12}b_{21} + a_{13}b_{31} & a_{11}b_{12} + a_{12}b_{22} + a_{13}b_{32} & a_{11}b_{13} + a_{12}b_{23} + a_{13}b_{33} \\ a_{21}b_{11} + a_{22}b_{21} + a_{23}b_{31} & a_{21}b_{12} + a_{22}b_{22} + a_{23}b_{32} & a_{21}b_{13} + a_{22}b_{23} + a_{23}b_{33} \\ a_{31}b_{11} + a_{32}b_{21} + a_{33}b_{31} & a_{31}b_{12} + a_{32}b_{22} + a_{33}b_{32} & a_{31}b_{13} + a_{32}b_{23} + a_{33}b_{33} \end{pmatrix} .$$

Thus we define this matrix to be the *product* of the matrix of **A** and the matrix of **B**.

Note that the multiplication of transformations and matrices is not necessarily commutative, i.e., **AB** is not necessarily the same as **BA**. It is left to the reader to show that, for example, if

$$\text{matrix } \mathbf{A} = \begin{pmatrix} 0 & 2 \\ 1 & 1 \end{pmatrix} \quad \text{and} \quad \text{matrix } \mathbf{B} = \begin{pmatrix} 2 & 2 \\ 0 & -1 \end{pmatrix},$$

then $\mathbf{AB} \neq \mathbf{BA}$.

Note that **I** and **O** commute with any transformation. That is, $\mathbf{IA} = \mathbf{AI}$ and $\mathbf{OA} = \mathbf{AO}$ for any **A**. We leave it to the reader to verify this by means of transformations and matrices.

We observe that the matrix of the transformation $k\mathbf{A}$ is obtained by multiplying every element in the matrix of **A** by k. We define this matrix to be the *product* of the scalar k and the matrix of **A**.

2.8 **Rectangular matrices:** In general a matrix is defined to be a rectangular array of scalars in which the number of rows is not necessarily equal to the number of columns. It is useful to consider some properties of such rectangular matrices as well as the square ones which were related to transformations. For example, it is sometimes convenient to treat a vector as a matrix of one row or one column.

Two rectangular matrices can not always be added or multiplied. In this section we are only interested in multiplication of matrices. If the number of columns in the first matrix is the same as the number of rows in the second, the multiplication is possible, and is defined exactly as in the case of square matrices.

Note that the inner product of two vectors (x_1, y_1, z_1) and (x_2, y_2, z_2) can be written as

$$(x_1x_2 + y_1y_2 + z_1z_2) = (x_1 \; y_1 \; z_1) \begin{pmatrix} x_2 \\ y_2 \\ z_2 \end{pmatrix}.$$

2.9 **Transform of a vector:** Let V be a vector in the space in which the base $\{U_1, U_2, U_3\}$ has been chosen. Let **A** be a linear transformation on the space such that its matrix with respect to $\{U_1, U_2, U_3\}$ is

$$\begin{pmatrix} a_{11} & a_{12} & a_{13} \\ a_{21} & a_{22} & a_{23} \\ a_{31} & a_{32} & a_{33} \end{pmatrix}$$

Suppose x, y, z are the components of V and X, Y, Z the components of $\mathbf{A}V$. For convenience we use the notation $V = (x, y, z)$ and $\mathbf{A}V = (X, Y, Z)$.

By 2.4 we know

$$\mathbf{A}U_1 = a_{11}U_1 + a_{12}U_2 + a_{13}U_3,$$
$$\mathbf{A}U_2 = a_{21}U_1 + a_{22}U_2 + a_{23}U_3,$$
$$\mathbf{A}U_3 = a_{31}U_1 + a_{32}U_2 + a_{33}U_3.$$

But by 1.7

$$V = xU_1 + yU_2 + zU_3, \text{ and}$$
$$\mathbf{A}V = XU_1 + YU_2 + ZU_3.$$

$$\mathbf{A}V = \mathbf{A}(xU_1 + yU_2 + zU_3) =$$
$$x(a_{11}U_1 + a_{12}U_2 + a_{13}U_3) + y(a_{21}U_1 + a_{22}U_2 + a_{23}U_3) + z(a_{31}U_1 + a_{32}U_2 + a_{33}U_3)$$
$$= (a_{11}x + a_{21}y + a_{31}z)\,U_1 + (a_{12}x + a_{22}y + a_{32}z)\,U_2 + (a_{13}x + a_{23}y + a_{33}z)\,U_3.$$

Thus

$$X = a_{11}x + a_{21}y + a_{31}z, \quad Y = a_{12}x + a_{22}y + a_{32}z, \quad Z = a_{13}x + a_{23}y + a_{33}z.$$

We observe that if (X, Y, Z) is considered as a row matrix we can show the effect of linear transformation as matrix multiplication as follows:

$$(1) \qquad \mathbf{A}V = (X\ Y\ Z) = (x\ y\ z)\begin{pmatrix} a_{11} & a_{12} & a_{13} \\ a_{21} & a_{22} & a_{23} \\ a_{31} & a_{32} & a_{33} \end{pmatrix} =$$

$$(a_{11}x + a_{21}y + a_{31}z \quad a_{12}x + a_{22}y + a_{32}z \quad a_{13}x + a_{23}y + a_{33}z)\ .$$

We also can use column matrices as

$$(2) \qquad \mathbf{A}V = \begin{pmatrix} X \\ Y \\ Z \end{pmatrix} = \begin{pmatrix} a_{11} & a_{21} & a_{31} \\ a_{12} & a_{22} & a_{32} \\ a_{13} & a_{23} & a_{33} \end{pmatrix}\begin{pmatrix} x \\ y \\ z \end{pmatrix} = \begin{pmatrix} a_{11}x + a_{21}y + a_{31}z \\ a_{12}x + a_{22}y + a_{32}z \\ a_{13}x + a_{23}y + a_{33}z \end{pmatrix}\ .$$

Note that the matrix used in (2) is the matrix in (1) with rows and columns interchanged. This matrix is called the *transpose* of the original matrix.

We call the system of equations

$$\begin{cases} X = a_{11}x + a_{21}y + a_{31}z \\ Y = a_{12}x + a_{22}y + a_{23}z \\ Z = a_{13}x + a_{23}y + a_{33}z \end{cases}$$

the *equations* of the transformation **A**.

Illustration 1: Find the product of the matrices

$$\begin{pmatrix} 3 & 1 \\ -2 & 0 \\ 2 & -1 \end{pmatrix},\quad \begin{pmatrix} 1 & -2 \\ -1 & 3 \end{pmatrix}$$

$$\begin{pmatrix} 3 & 1 \\ -2 & 0 \\ 2 & -1 \end{pmatrix}\begin{pmatrix} 1 & -2 \\ -1 & 3 \end{pmatrix} = \begin{pmatrix} (3)(1) + (1)(-1) & (3)(-2) + (1)(3) \\ (-2)(1) + (0)(-1) & (-2)(-2) + (0)(3) \\ (2)(1) + (-1)(-1) & (2)(-2) + (-1)(3) \end{pmatrix} = \begin{pmatrix} 2 & -3 \\ -2 & 4 \\ 3 & -7 \end{pmatrix}.$$

Note that the product

$$\begin{pmatrix} 1 & -2 \\ -1 & 3 \end{pmatrix}\begin{pmatrix} 3 & 1 \\ -2 & 0 \\ 2 & -1 \end{pmatrix}$$

is not defined.

Illustration 2: Let the matrix of **A** with respect to some base be

$$\begin{pmatrix} 2 & -1 & 0 \\ -2 & -7 & 1 \\ 1 & 1 & 3 \end{pmatrix}.$$

If $V = (2,-1,3)$, find $\mathbf{A}V$.

$$\mathbf{A}V = (2 \quad -1 \quad 3)\begin{pmatrix} 2 & -1 & 0 \\ -2 & -7 & 1 \\ 1 & 1 & 3 \end{pmatrix} = (9 \quad 8 \quad 8), \text{ or}$$

$$\mathbf{A}V = \begin{pmatrix} 2 & -2 & 1 \\ -1 & -7 & 1 \\ 0 & 1 & 3 \end{pmatrix}\begin{pmatrix} 2 \\ -1 \\ 3 \end{pmatrix} = \begin{pmatrix} 9 \\ 8 \\ 8 \end{pmatrix}.$$

Illustration 3: Let the matrix of $\mathbf{A}$ with respect to some base be

$$\begin{pmatrix} 2 & -1 & 0 \\ -2 & -7 & 1 \\ 1 & 1 & 3 \end{pmatrix}.$$

Find the equations of the transformation.
Let $V = (x,y,z)$, $\mathbf{A}V = (X,Y,Z)$. Then

$$(X \ Y \ Z) = \mathbf{A}V = (x \ y \ z)\begin{pmatrix} 2 & -1 & 0 \\ -2 & -7 & 1 \\ 1 & 1 & 3 \end{pmatrix} = (2x-2y+z \quad -x-7y+z \quad y+3z),$$

thus

$$\begin{cases} X = 2x - 2y + z \\ Y = -x - 7y + z \\ Z = \quad\quad y + 3z . \end{cases}$$

EXERCISES 2

1. Discuss the geometric meaning of the following linear transformations in which (x,y) is transformed to (x',y').

   (i) $\begin{cases} y' = x \\ x' = y \end{cases}$, (ii) $\begin{cases} y' = x \\ x' = 0 \end{cases}$,

   (iii) $\begin{cases} y' = 2y \\ x' = 3x \end{cases}$, (iv) $\begin{cases} x' = x + y \\ y' = x - y \end{cases}$.

2. Write the matrix of each of the transformations in 1.
3. Discuss the geometric meaning of the following transformations where (x,y,z) goes to (x', y', z').

   (i) $\begin{cases} x' = 2x \\ y' = 3y \\ z' = 5z \end{cases}$, (ii) $\begin{cases} x' = 0 \\ y' = 0 \\ z' = z \end{cases}$, (iii) $\begin{cases} x' = x + 2y \\ y' = z \\ z' = 2y \end{cases}$.

4. Write the matrix of each of the transformations in 3.
5. Write the matrix of

   (i) $\begin{cases} \mathbf{A}(2,2) = (0,1) \\ \mathbf{A}(-1,0) = (3,2) \end{cases}$, (ii) $\begin{cases} \mathbf{A}(2,3) = (1,0) \\ \mathbf{A}(3,0) = (1,-1) \end{cases}$,

(iii) $\begin{cases} \mathbf{A}(1,0,0) = (1,2,1) \\ \mathbf{A}(0,1,0) = (3,1,1) \\ \mathbf{A}(0,0,1) = (0,0,3) \end{cases}$, (iv) $\begin{cases} \mathbf{A}(2,0,0) = (5,1,1) \\ \mathbf{A}(-1,-1,0) = (0,0,1) \\ \mathbf{A}(0,0,1) = (1,-1,0) \end{cases}$.

6. If

$$\text{matrix } \mathbf{A} = \begin{pmatrix} 1 & 1 \\ 0 & 1 \end{pmatrix}, \quad \text{matrix } \mathbf{B} = \begin{pmatrix} 1 & -1 \\ 0 & 1 \end{pmatrix}, \quad \text{matrix } \mathbf{C} = \begin{pmatrix} 1 & 3 \\ 2 & 1 \end{pmatrix},$$

compute the matrices

(i) $\mathbf{AB}$, $\mathbf{BA}$, $\mathbf{A}^2$, $\mathbf{B}^3$, $3\mathbf{A} + \mathbf{B}$, and $\mathbf{A} - 2\mathbf{C}$;

(ii) $(\mathbf{A} - \mathbf{I})(\mathbf{I} + \mathbf{C})(\mathbf{C} - \mathbf{A})$.

In the following exercises, use a convenient orthonormal base.

7. Find the matrix of a rotation of the plane about the origin through an angle of

(i) $30°$, (ii) $45°$, (iii) arbitrary θ .

8. Let $\mathbf{A}$ be the rotation of the plane through an angle of $30°$, $\mathbf{B}$ the rotation through $45°$. Find the equations of each of these transformations.
9. Let $\mathbf{C} = \mathbf{AB}$. From the equations of $\mathbf{A}$ and $\mathbf{B}$ in exercise 8, find the equations of $\mathbf{C}$ by substitution, and the matrix of $\mathbf{C}$.
10. Find the matrix of $\mathbf{C}$ by matrix multiplication, using the result of exercise 7. Compare with the matrix found in exercise 9.
11. Show that, for the transformations $\mathbf{A}$ and $\mathbf{B}$ of exercise 8, $\mathbf{AB} = \mathbf{BA}$, and show that both represent the rotation through $75°$.
12. Let $\mathbf{A}$ and $\mathbf{B}$ be rotations of the plane through angles θ and ϕ respectively. Find the matrix of $\mathbf{C} = \mathbf{AB}$ by matrix multiplication and by composition of equations of transformations, and compare the results.
13. For the transformations $\mathbf{A}$ and $\mathbf{B}$ of exercise 12, show that $\mathbf{AB} = \mathbf{BA}$, and interpret this product geometrically.
14. Find the matrix and the equations of the symmetry with respect to

(i) the origin ,

(ii) the x-axis ,

(iii) the y-axis ,

(iv) the line $y = x$,

(v) the line $ax + by = 0$.

15. Find the matrix and the equations of the projection on

(i) the x-axis ,

(ii) the line $y = 3x$,

(iii) the line $cx + dy = 0$.

16. Let $\mathbf{A}$ by any one of the symmetries of exercise 14. Let $\mathbf{B}$ be a projection of exercise 15. In all cases find $\mathbf{AB}$ and $\mathbf{BA}$.

17. Find the matrix and the equations of the symmetry with respect to:

(i) the origin, (ii) the x-axis, (iii) the y-axis,

(iv) the z-axis, (v) $z = 0$, (vi) $y = 0$,

(vii) $x = 0$, (viii) $x + y + z = 0$, (ix) $ax + by + cz = 0$.

18. Find the matrix and the equations of the rotation through an angle θ about

(i) the x-axis , (ii) the z-axis ,

(iii) the line $x = y = z$, (iv) the line $ax = by = cz$.

19. Find the matrix and the equations of the projection on

(i) the xy-plane , (ii) the x-axis ,

(iii) the plane $x + y + z = 0$, (iv) the line $x = y = z$,

(v) the plane $ax + by + cz = 0$.

20. Find the product of various transformations in exercises 17, 18, and 19.

ADDITIONAL PROBLEMS 2

In these problems choose orthonormal bases.

1. Find the matrix and equations of a rotation of the plane about the origin through an angle

(i) $60°$, (ii) $90°$, (iii) $120°$,

(iv) $150°$, (v) $210°$, (vi) $240°$.

2. The same as problem 1 for

(i) $-30°$, (ii) $-60°$,

(iii) $\frac{5\pi}{6}$, (iv) $\frac{7\pi}{6}$,

3. The same as problem 1 for

(i) $135°$, (ii) $-45°$,

(iii) $-135°$, (iv) $225°$.

4. Use a convenient matrix multiplication to find the matrix and equations of a rotation of the plane through an angle

(i) $105°$, (ii) $165°$,

(iii) $195°$, (iv) $255°$.

5. From the matrix of a rotation of the plane about the origin through an angle $75°$ find the matrix of a rotation of the plane about the origin through an angle $150°$.
6. Show that if we square the matrix of a rotation of the plane about the origin through an angle θ we get the matrix of rotation of the plane about the origin through an angle 2θ.
7. State and prove the equivalent of problem 5 for any angle $n\theta$ where n is an integer.
8. Let $\mathbf{A}_1, \mathbf{A}_2, \ldots, \mathbf{A}_k$ be k rotations of the plane about the origin through angles $\theta_1, \theta_2, \ldots, \theta_k$. Show that

(i) $\mathbf{A}_1\mathbf{A}_2 \cdot \ldots \cdot \mathbf{A}_k$ is a rotation through an angle $\theta_1 + \theta_2 + \ldots + \theta_k$.

(ii) The multiplication of these matrices is commutative.

9. Find the matrix and equations of a rotation of the space about the x-axis through an angle

(i) $60°$, (ii) $120°$, (iv) $150°$,

(iv) $240°$, (v) $135°$, (vi) $-45°$.

10. Find the matrix and equations of a rotation about the y-axis through an angle

(i) $\frac{5\pi}{6}$, (ii) $\frac{7\pi}{6}$,

(iii) $-135°$, (iv) $165°$.

11. Use a convenient matrix multiplication to find the matrix and equations of a rotation about the line $x = y = z$ through an angle

(i) $165°$, (ii) $105°$.

12. Let a transformation **A** be given by

$$\begin{cases} \mathbf{A}(3,2,1) = (0,0,2) \\ \mathbf{A}(0,1,1) = (1,1,1) \\ \mathbf{A}(2,0,1) = (1,0,1) . \end{cases}$$

Find the matrix of **A** with respect to the base $\{(1,0,0), (0,1,0), (0,0,1)\}$.

13. Let **A** be a permutation of the vectors of the base. Find the matrices of all possible such permutations.
14. Let **A** be the symmetry with respect to the x-axis in the plane and **B** be the symmetry with respect to the y-axis. Show that **AB** = **BA** and **AB** is the symmetry with respect to the origin.
15. Let **A** be the symmetry with respect to the plane $x = 0$, **B** the symmetry with respect to $y = 0$, and **C** the symmetry with respect to $z = 0$. Show that **ABC** is the symmetry with respect to the origin. Also show that the product of **A**, **B**, and **C** is independent of the order of multiplication.
16. Show that, in the plane, the symmetry with respect to the origin is the same as a rotation through an angle π. Besides a geometric proof supply the matrix proof.
17. Let **A** be a symmetry of the plane with respect to the x-axis, and **B** the symmetry with respect to $y = x$. Show that **AB** $\neq$ **BA**.
18. Let **A** be a rotation about the z-axis through an angle α, and **B** be a rotation about the y-axis through an angle β. Show that **AB** $\neq$ **BA**.
19. Show that the transformation whose equations are

$$\begin{cases} X = ax + by + c \\ Y = dx + fy + g \end{cases}$$

is not in general linear. What is the condition for which the transformation is linear?
20. Show that the translation is not a linear transformation.
21. Let **A** be the symmetry with respect to $y = kx$ and **B** be the symmetry with respect to $x = -ky$. Show that **AB** = **BA**. Besides a geometric proof supply the matrix proof.
22. If **A** is the symmetry with respect to $y = mx$ and **B** the symmetry with respect to $y = nx$, $m \neq n$. Show that in general **AB** $\neq$ **BA**. Also show that $mn = -1$ is the condition for **AB** = **BA**. Geometric and matrix proofs both are expected.
23. Let **A** be the projection in the plane on $y = mx$ and **B** the projection on $x = -my$. Show that **A** + **B** = **I**.
24. Let **A**, **B**, and **C** be projections on

$\frac{x}{3} = \frac{y}{2} = -\frac{z}{6}$, $-\frac{x}{6} = \frac{y}{3} = -\frac{z}{2}$, and $\frac{x}{2} = \frac{y}{6} = \frac{z}{3}$ respectively. Show

that

(i) $\mathbf{A}^2 = \mathbf{A}$, $\mathbf{B}^2 = \mathbf{B}$, and $\mathbf{C}^2 = \mathbf{C}$;

(ii) $\mathbf{A} + \mathbf{B} + \mathbf{C} = \mathbf{I}$.

25. Let $\mathbf{A}$ be the projection on the plane $ax + by + cz = 0$, and $\mathbf{B}$ be the projection on $\frac{x}{a} = \frac{y}{b} = \frac{z}{c}$. Show that

(i) $\mathbf{A}^2 = \mathbf{A}$, (ii) $\mathbf{A} + \mathbf{B} = \mathbf{I}$.

26. Let $\mathbf{A}$ be the symmetry with respect to $ax + by + cz = 0$. Show that $\mathbf{A}^2 = \mathbf{I}$.

27. Show that if $\mathbf{A}$ is any symmetry, then $\mathbf{A}^2 = \mathbf{I}$.

3. DETERMINANTS AND LINEAR EQUATIONS

3.1 **Definition:** For a one-by-one matrix (a) the *determinant* of (a) is the scalar a. The determinant of

$$\begin{pmatrix} a_{11} & a_{12} \\ a_{21} & a_{22} \end{pmatrix}, \text{ denoted by } \begin{vmatrix} a_{11} & a_{12} \\ a_{21} & a_{22} \end{vmatrix},$$

is defined to be $a_{11}a_{22} - a_{12}a_{21}$.

For a three-by-three matrix the determinant is again denoted by

$$\begin{vmatrix} a_{11} & a_{12} & a_{13} \\ a_{21} & a_{22} & a_{23} \\ a_{31} & a_{32} & a_{33} \end{vmatrix}.$$

Given any term a_{ij}, i.e., the term in the i-th row and the j-th column, we define the *cofactor* of a_{ij} to be $\mathbf{A}_{ij}$ where

$$\mathbf{A}_{ij} = (-1)^{i+j} \begin{vmatrix} h & k \\ l & m \end{vmatrix}, \text{ and } \begin{vmatrix} h & k \\ l & m \end{vmatrix}$$

is the determinant obtained from $\begin{vmatrix} a_{11} & a_{12} & a_{13} \\ a_{21} & a_{22} & a_{23} \\ a_{31} & a_{32} & a_{33} \end{vmatrix}$

by removing the i-th row and the j-th column.

For example

$$\mathbf{A}_{23} = (-1)^{2+3} \begin{vmatrix} a_{11} & a_{12} \\ a_{31} & a_{32} \end{vmatrix}$$

Now the determinant of $\begin{pmatrix} a_{11} & a_{12} & a_{13} \\ a_{21} & a_{22} & a_{23} \\ a_{31} & a_{32} & a_{33} \end{pmatrix}$ is

defined as follows:

$$\begin{vmatrix} a_{11} & a_{12} & a_{13} \\ a_{21} & a_{22} & a_{23} \\ a_{31} & a_{32} & a_{33} \end{vmatrix} = a_{1j}\mathbf{A}_{1j} + a_{2j}\mathbf{A}_{2j} + a_{3j}\mathbf{A}_{3j}, \text{ or } a_{i1}\mathbf{A}_{i1} + a_{i2}\mathbf{A}_{i2} + a_{i3}\mathbf{A}_{i3}.$$

It is necessary to show that the result is independent of the choice of i or j. The reader can show by elementary algebra that in any case the value of the determinant will be

(1) $$a_{11}a_{22}a_{33} - a_{11}a_{32}a_{23} - a_{21}a_{12}a_{33} + a_{21}a_{32}a_{13} + a_{31}a_{12}a_{23} - a_{31}a_{22}a_{13}$$

3.2 **Some properties of determinants:** We observe that the product terms of 3.1 (1) each involve one term of each row and each column. Thus:

(1) if a row or a column of a determinant consists of zeros, the value of the determinant is zero;
(2) if the terms of any one row or column are multiplied by a scalar, the value of the determinant is multiplied by the same scalar.
(3) It is a simple algebraic exercise to verify that if two rows or two columns of a two-by-two or three-by-three matrix are interchanged, the determinant of the resulting matrix is the negative of the original.
(4) Hence if a matrix has two of its rows or two of its columns identical, the value of the determinant is zero.
(5) Thus if the terms of any row are multiplied by the cofactors of another row the sum of the resulting products is zero.
(6) It may be shown that if two matrices are identical except for one row, then for example

$$\begin{vmatrix} a_{11} & a_{12} & a_{13} \\ a_{21} & a_{22} & a_{23} \\ a_{31} & a_{32} & a_{33} \end{vmatrix} + \begin{vmatrix} b_{11} & b_{12} & b_{13} \\ a_{21} & a_{22} & a_{23} \\ a_{31} & a_{32} & a_{33} \end{vmatrix} = \begin{vmatrix} a_{11}+b_{11} & a_{12}+b_{12} & a_{13}+b_{13} \\ a_{21} & a_{22} & a_{23} \\ a_{31} & a_{32} & a_{33} \end{vmatrix} .$$

(7) Thus if a row of a matrix is altered by adding to it a scalar multiple of some other row, the value of its determinant is unchanged.
(8) Hence if we consider each row of a matrix as a vector, and if one row vector of the matrix is a linear combination of the others, the value of the determinant is zero. Obviously the statements of (5), (6), (7), and (8) are true for columns as well as rows, and the determinant of the transpose of a matrix is equal to the determinant of the matrix.
(9) We observe that the determinant of the matrix of the identity transformation is equal to 1.

3.3 **Theorem:** The determinant of the product of two matrices is equal to the product of the determinants of the two matrices.

Proof: The theorem is trivial for one-by-one matrices. We prove the theorem for two-by-two matrices and we leave the case of three-by-three as an exercise.

$$\begin{pmatrix} a_{11} & a_{12} \\ a_{21} & a_{22} \end{pmatrix} \begin{pmatrix} b_{11} & b_{12} \\ b_{21} & b_{22} \end{pmatrix} = \begin{pmatrix} a_{11}b_{11}+a_{12}b_{21} & a_{11}b_{12}+a_{12}b_{22} \\ a_{21}b_{11}+a_{22}b_{21} & a_{21}b_{12}+a_{22}b_{22} \end{pmatrix}$$

Therefore

$$\begin{vmatrix} a_{11}b_{11}+a_{12}b_{21} & a_{11}b_{12}+a_{12}b_{22} \\ a_{21}b_{11}+a_{22}b_{21} & a_{21}b_{12}+a_{22}b_{22} \end{vmatrix} = (a_{11}b_{11}+a_{12}b_{21})(a_{21}b_{12}+a_{22}b_{22})-(a_{11}b_{12}+a_{12}b_{22})(a_{21}b_{11}+a_{22}b_{21})$$

$$= (a_{11}a_{22}-a_{12}a_{21})(b_{11}b_{22}-b_{12}b_{21}) = \begin{vmatrix} a_{11} & a_{12} \\ a_{21} & a_{22} \end{vmatrix} \cdot \begin{vmatrix} b_{11} & b_{12} \\ b_{21} & b_{22} \end{vmatrix} .$$

3.4 **Systems of linear equations:** If we are given a pair of equations in two unknowns,

$$(1) \qquad \begin{cases} a_{11}x + a_{12}y = b_1 \\ a_{21}x + a_{22}y = b_2 , \end{cases}$$

let us determine when there are solutions, and then what solutions exist. If we multiply the first equation by a_{22}, the second by a_{12} and subtract the second from the first, then multiply the first by a_{21}, the second by a_{11}, and subtract in the other order we get

$$\begin{cases} (a_{11}a_{22} - a_{21}a_{12})x = b_1 a_{22} - b_2 a_{12}, \\ (a_{11}a_{22} - a_{21}a_{12})y = a_{11}b_2 - a_{21}b_1, \end{cases} \quad \text{that is,}$$

$$(2) \qquad \begin{cases} \begin{vmatrix} a_{11} & a_{12} \\ a_{21} & a_{22} \end{vmatrix} x = \begin{vmatrix} b_1 & a_{12} \\ b_2 & a_{22} \end{vmatrix} \\ \begin{vmatrix} a_{11} & a_{12} \\ a_{21} & a_{22} \end{vmatrix} y = \begin{vmatrix} a_{11} & b_1 \\ a_{21} & b_2 \end{vmatrix} \end{cases} .$$

Thus if the determinant of the coefficients of x and y is not equal to zero, there exists a unique solution of (1) for any choice of b_1 and b_2. In particular if $b_1 = b_2 = 0$, then the only solution of (1) is $x = y = 0$. We refer to the system (1) in which $b_1 = b_2 = 0$ as *homogenous*.

If however $\begin{vmatrix} a_{11} & a_{12} \\ a_{21} & a_{22} \end{vmatrix} = 0$, we must investigate further.

We may assume that at least one of the numbers a_{ij}, say a_{11} is not zero. Then for any choice of y, we have, from the first equation of (1),

$$x = \frac{b_1 - a_{12}y}{a_{11}} .$$

This will give a solution if we have also

$$a_{21}\frac{b_1 - a_{12}y}{a_{11}} + a_{22}y = b_2 ,$$

or $\qquad (a_{11}a_{22} - a_{21}a_{12})y = a_{11}b_2 - a_{21}b_1 , \qquad$ whence, since

$$a_{11}a_{22} - a_{21}a_{12} = 0, \text{ we have } a_{11}b_2 - a_{21}b_1 = \begin{vmatrix} a_{11} & b_1 \\ a_{21} & b_2 \end{vmatrix} = 0 .$$

Since in this case the pair x, y is a solution of (1), by (2), it must also be true that $\begin{vmatrix} b_1 & a_{12} \\ b_2 & a_{22} \end{vmatrix} = 0$.

Thus when the determinant of coefficients is equal to zero, solutions of (1) exist if and only if

$$\begin{vmatrix} b_1 & a_{12} \\ b_2 & a_{22} \end{vmatrix} = \begin{vmatrix} a_{11} & b_1 \\ a_{21} & a_2 \end{vmatrix} = 0 .$$

Finally we observe that if (1) is a homogeneous system, solutions other than $x = y = 0$ will exist if and only if $\begin{vmatrix} a_{11} & a_{12} \\ a_{21} & a_{22} \end{vmatrix} = 0$.

We consider now the system

$$(3)\quad \begin{cases} a_{11}x + a_{12}y + a_{13}z = b_1 \\ a_{21}x + a_{22}y + a_{23}z = b_2 \\ a_{31}x + a_{32}y + a_{33}z = b_3 \ . \end{cases}$$

If we multiply the first of the equations of (3) by $\mathbf{A}_{11}$, the second by $\mathbf{A}_{21}$, and the third by $\mathbf{A}_{31}$, and add, we eliminate the terms y and z. [see 3.1 and 3.2 (5)].

We get

$$(a_{11}\mathbf{A}_{11} + a_{21}\mathbf{A}_{21} + a_{31}\mathbf{A}_{31})x = b_1\mathbf{A}_{11} + b_2\mathbf{A}_{21} + b_3\mathbf{A}_{31}, \text{ or}$$

$$(4a)\quad \begin{vmatrix} a_{11} & a_{12} & a_{13} \\ a_{21} & a_{22} & a_{23} \\ a_{31} & a_{32} & a_{33} \end{vmatrix} x = \begin{vmatrix} b_1 & a_{12} & a_{13} \\ b_2 & a_{22} & a_{23} \\ b_3 & a_{32} & a_{33} \end{vmatrix} .$$

In similar fashion, multiplying by $\mathbf{A}_{12}$, $\mathbf{A}_{22}$, $\mathbf{A}_{32}$ and then by $\mathbf{A}_{13}$, $\mathbf{A}_{23}$, $\mathbf{A}_{33}$ we get

$$(4b)\quad \begin{vmatrix} a_{11} & a_{12} & a_{13} \\ a_{21} & a_{22} & a_{23} \\ a_{31} & a_{32} & a_{33} \end{vmatrix} y = \begin{vmatrix} a_{11} & b_1 & a_{13} \\ a_{21} & b_2 & a_{23} \\ a_{31} & b_3 & a_{33} \end{vmatrix} ,$$

$$(4c)\quad \begin{vmatrix} a_{11} & a_{12} & a_{13} \\ a_{21} & a_{22} & a_{23} \\ a_{31} & a_{32} & a_{33} \end{vmatrix} z = \begin{vmatrix} a_{11} & a_{12} & b_1 \\ a_{21} & a_{22} & b_2 \\ a_{31} & a_{32} & b_3 \end{vmatrix} .$$

Again if $\begin{vmatrix} a_{11} & a_{12} & a_{13} \\ a_{21} & a_{22} & a_{23} \\ a_{31} & a_{32} & a_{32} \end{vmatrix} \neq 0,$ there is a unique solution of the system (3), and if the equations are homogeneous, i.e. $b_1 = b_2 = b_3 = 0$, this solution is $x = y = z = 0$.

If the determinant of the coefficients is equal to zero, again solutions may exist if the right members of the equations (4) are all equal to zero. Let us assume that some two-by-two determinant in the coefficient matrix is not zero, say for example

$$\begin{vmatrix} a_{11} & a_{12} \\ a_{21} & a_{22} \end{vmatrix} = 0 \ .$$

We write (3) in the form

$$\begin{cases} a_{11}x + a_{12}y = b_1 - a_{13}z \\ a_{21}x + a_{22}y = b_2 - a_{23}z \\ a_{31}x + a_{32}y + b_3 - a_{33}z \ . \end{cases}$$

Then for any choice of z, the first two of these equations have the unique solution

$$x = \frac{\begin{vmatrix} b_1 - a_{13}z & a_{12} \\ b_2 - a_{23}z & a_{22} \end{vmatrix}}{\begin{vmatrix} a_{11} & a_{12} \\ a_{21} & a_{22} \end{vmatrix}}$$

$$y = \frac{\begin{vmatrix} a_{11} & b_1 - a_{13}z \\ a_{21} & b_2 - a_{23}z \end{vmatrix}}{\begin{vmatrix} a_{11} & a_{12} \\ a_{21} & a_{22} \end{vmatrix}}$$

The triple (x, y, z) will be a solution of (3) if

$$a_{31}\begin{vmatrix} b_1 - a_{13}z & a_{12} \\ b_2 - a_{23}z & a_{22} \end{vmatrix} + a_{32}\begin{vmatrix} a_{11} & b_1 - a_{13}z \\ a_{21} & b_2 - a_{23}z \end{vmatrix} = b_3\begin{vmatrix} a_{11} & a_{12} \\ a_{21} & a_{22} \end{vmatrix} - a_{33}z\begin{vmatrix} a_{11} & a_{12} \\ a_{21} & a_{22} \end{vmatrix},$$

that is, if

$$\left\{ a_{31}\begin{vmatrix} a_{12} & a_{13} \\ a_{22} & a_{23} \end{vmatrix} - a_{32}\begin{vmatrix} a_{11} & a_{13} \\ a_{21} & a_{23} \end{vmatrix} + a_{33}\begin{vmatrix} a_{11} & a_{12} \\ a_{21} & a_{22} \end{vmatrix} \right\} z$$

$$= a_{31}\begin{vmatrix} a_{12} & b_1 \\ a_{22} & b_2 \end{vmatrix} - a_{32}\begin{vmatrix} a_{11} & b_1 \\ a_{21} & b_2 \end{vmatrix} + b_3\begin{vmatrix} a_{11} & a_{12} \\ a_{21} & a_{22} \end{vmatrix},$$

or

$$\begin{vmatrix} a_{11} & a_{12} & a_{13} \\ a_{21} & a_{22} & a_{23} \\ a_{31} & a_{32} & a_{33} \end{vmatrix} z = \begin{vmatrix} a_{11} & a_{12} & b_1 \\ a_{21} & a_{22} & b_2 \\ a_{31} & a_{32} & b_3 \end{vmatrix}$$

Thus solutions will exist if

$$\begin{vmatrix} a_{11} & a_{12} & b_1 \\ a_{21} & a_{22} & b_2 \\ a_{31} & a_{32} & b_3 \end{vmatrix} = 0 .$$

in which case the right members of equations (4) must all be equal to zero. In particular there will exist a non-zero solution for the homogeneous equations.

If every two-by-two determinant of the coefficient matrix is zero, we may again assume that at least one coefficient, say a_{11}, is not zero. From the first equation of (3), we must have, for any solution (x,y,z),

$$x = \frac{b_1 - a_{11}y - a_{13}z}{a_{11}}$$

For any choice of y and z this will be a solution if

$$a_{21}(b_1 - a_{12}y - a_{13}z) + a_{11}a_{22}y + a_{11}a_{23}z = a_{11}b_2\,, \text{ and}$$

$$a_{31}(b_1 - a_{12}y - a_{13}z) + a_{11}a_{32}y + a_{11}a_{33}z = a_{11}b_3\,,$$

that is, if

$$(a_{11}a_{22} - a_{21}a_{12})y + (a_{11}a_{23} - a_{21}a_{13})z = a_{11}b_2 - a_{21}b_1\,, \text{ and}$$

$$(a_{11}a_{32} - a_{31}a_{12})y + (a_{11}a_{33} - a_{31}a_{13})z = a_{11}b_3 - a_{31}b_1\,.$$

But by assumption every two-by-two determinant of the coefficient matrix is equal to zero, so that we must have

$$\begin{vmatrix} a_{11} & b_1 \\ a_{21} & b_2 \end{vmatrix} = \begin{vmatrix} a_{11} & b_1 \\ a_{31} & b_3 \end{vmatrix} = 0, \text{ whence also } \begin{vmatrix} a_{21} & b_2 \\ a_{31} & b_3 \end{vmatrix} = 0\,.$$

In these circumstances it can be shown that the three two-by-two determinants obtained from those above by replacing a_{i1} by a_{i2} and the three obtained by using a_{i3}, $i = 1, 2, 3$, are also equal to zero.

We have demonstrated in this case that solutions exist if and only if all of the two-by-two determinants containing a column of the constants b_i of (3) are equal to zero. In particular this will be true if every $b_i = 0$.

We call attention to the fact that for the homogeneous system, solutions other than $x = y = z = 0$ (non-trivial solutions) *will exist if and only if the determinant of the coefficients is zero.*

Illustration 1: Find the point of intersection of the lines

$$3x + 2y = 5 \quad \text{and} \quad 6x + 4y = 7\,.$$

A point of intersection of these lines is a solution of the system of equations

$$\begin{cases} 3x + 2y = 5 \\ 6x + 4y = 7\,. \end{cases}$$

We observe that

$$\begin{vmatrix} 3 & 2 \\ 6 & 4 \end{vmatrix} = 0, \text{ while } \begin{vmatrix} 5 & 2 \\ 7 & 4 \end{vmatrix} \neq 0\,,$$

whence the system has no solutions. Note that the two lines in this case are parallel, and we would have expected this result.

Illustration 2: Consider lines $a_1x + b_1y = c_1$ and $a_2x + b_2y = c_2$. Show that if the determinant of the coefficients of the system

$$\begin{cases} a_1x + b_1y = c_1 \\ a_2x + b_2y = c_2 \end{cases}$$

is zero, the lines are either parallel or coincident.

If $\begin{vmatrix} a_1 & b_1 \\ a_2 & b_2 \end{vmatrix} = 0$, then $a_1 b_2 = b_1 a_2$. If $b_1 \neq 0$, then $b_2 \neq 0$, since otherwise we would have $a_2 = 0$, and the second equation would not represent a line. In this case,

$$\frac{a_1}{b_1} = \frac{a_2}{b_2} .$$

If $b_1 = 0$, since $a_1 \neq 0$, similarly $a_2 \neq 0$ and

$$\frac{b_1}{a_1} = \frac{b_2}{a_2} .$$

We see that (a_1,b_1) and (a_2,b_2) are the directions perpendicular to the given lines. Thus $(a_2,b_2) = k(a_1,b_1)$, and the lines have the same direction. Now if $c_2 = kc_1$, then the two equations represent the same line. Otherwise the lines are parallel.

Illustration 3: Show that

$$\begin{vmatrix} x & y & 1 \\ -2 & 1 & 1 \\ 3 & 2 & 1 \end{vmatrix} = 0$$

is an equation of the line through the points (–2,1) and (3,2).

Let $ax + by + c = 0$ be the equation of the desired line. Then non-trivial numbers a, b, and c exist satisfying the system

$$\begin{cases} xa + yb + c = 0 \\ -2a + b + c = 0 \\ 3a + 2b + c = 0 . \end{cases}$$

This homogeneous system will have solutions other than (0,0,0) if the determinant of the coefficients is zero, that is, if

$$\begin{vmatrix} x & y & 1 \\ -2 & 1 & 1 \\ 3 & 2 & 1 \end{vmatrix} = 0 .$$

As an alternative solution, we may note that if we expand this determinant in terms of its first row, we get an equation of first degree in x and y. Since the pairs (–2,1) and (3,2) obviously satisfy the equation (the determinant then having two equal rows), this is an equation for the line through those points.

EXERCISES 3

1. Evaluate:

(i) $\begin{vmatrix} 3 & -7 \\ 1 & -2 \end{vmatrix}$, (ii) $\begin{vmatrix} 7 & 0 & 5 \\ 1 & -2 & 9 \\ 8 & -2 & 14 \end{vmatrix}$,

(iii) $\begin{vmatrix} 2 & 1 & -1 \\ 3 & 4 & 2 \\ 5 & 0 & 2 \end{vmatrix}$, (iv) $\begin{vmatrix} x & y & z \\ r & s & t \\ ax+br & ay+bs & az+bt \end{vmatrix}$.

2. Show that (Vandermonde)

$$\begin{vmatrix} 1 & x & x^2 \\ 1 & y & y^2 \\ 1 & z & z^2 \end{vmatrix} = (x-y)\,(y-z)\,(z-x)\ .$$

3. Show that

$$\begin{vmatrix} x & y & 1 \\ x_1 & y_1 & 1 \\ x_2 & y_2 & 1 \end{vmatrix} = 0$$

is the equation of the line passing through (x_1,y_1) and (x_2,y_2).

4. Show that the area of the triangle with vertices (x_1,y_1), (x_2,y_2), (x_3,y_3) is the absolute value of

$$\frac{1}{2}\begin{vmatrix} x_1 & y_1 & 1 \\ x_2 & y_2 & 1 \\ x_3 & y_3 & 1 \end{vmatrix}\ .$$

5.* Show that the area of the triangle with vertices (0,0,0), (x_1,y_1,z_1), and (x_2,y_2,z_2) is one half of the square root of the determinant of $\mathbf{AA}'$, where

$$\mathbf{A} = \begin{pmatrix} x_1 & y_1 & z_1 \\ x_2 & y_2 & z_2 \end{pmatrix}\ ,$$ and $\mathbf{A}'$ is the transpose of $\mathbf{A}$ (see 2.10).

6. Find the determinants of the transformations in chapter 2, exercises 2, 4, 5.
7. Find the determinants in chapter 2 exercises 6. Verify that the determinants of the products are the products of the determinants.
8. Show that any projection on a subspace has determinant equal to zero.
9. Let $\mathbf{A}$ be the symmetry with respect to $ax + by + cz = 0$. Find the determinants of $\mathbf{A}$ and $\mathbf{A}^2$.
10. Evaluate

(i) $\begin{vmatrix} \frac{3}{5} & \frac{4}{5} \\ -\frac{4}{5} & \frac{3}{5} \end{vmatrix}$ (ii) $\begin{vmatrix} \frac{\sqrt{3}}{2} & \frac{-1}{2} \\ \frac{1}{2} & \frac{\sqrt{3}}{2} \end{vmatrix}$,

(iii) $\begin{vmatrix} \frac{-2}{3} & \frac{1}{3} & \frac{2}{3} \\ \frac{-1}{3} & \frac{2}{3} & \frac{-2}{3} \\ \frac{-2}{3} & \frac{-2}{3} & \frac{-1}{3} \end{vmatrix}$, (iv) $\begin{vmatrix} \frac{3}{7} & \frac{2}{3} & \frac{-6}{7} \\ \frac{-6}{7} & \frac{3}{7} & \frac{-2}{7} \\ \frac{2}{7} & \frac{6}{7} & \frac{3}{7} \end{vmatrix}$.

11. Evaluate

(i) $\begin{vmatrix} \cos\theta & \sin\theta \\ -\sin\theta & \cos\theta \end{vmatrix}$, (ii) $\begin{vmatrix} \cos\theta & 0 & \sin\theta \\ 0 & 1 & 0 \\ -\sin\theta & 0 & -\cos\theta \end{vmatrix}$, (iii) $\begin{vmatrix} \sin\theta\cos\phi & \sin\theta\sin\phi & \cos\theta \\ \cos\theta\cos\phi & \cos\theta\sin\phi & -\sin\theta \\ \sin\phi & -\cos\phi & 0 \end{vmatrix}$.

12. Solve using determinants

(i) $\begin{cases} 3x + 5y = 1 \\ -x - 3y = 1 , \end{cases}$ (ii) $\begin{cases} x + y + z = 6 \\ 2x + y - z = 1 \\ 2y + 3z = 13 , \end{cases}$ (iii) $\begin{cases} 2x + y = 2 \\ 3y - 2z = 4 \\ y + 3z = 1 . \end{cases}$

13. Find all solutions, if any exist

(i) $\begin{cases} x + y + z = 6 \\ 3x - 2y - z = 7 \\ x - 4y - 3z = -5 , \end{cases}$ (ii) $\begin{cases} x + 2y - z = 5 \\ 3x - y + 2z = 2 \\ 2x + 11y - 7z = -2 , \end{cases}$ (iii) $\begin{cases} 3x + y - z = 0 \\ 2x + 2y - 3z = 0 \\ x - 5y + 9z = 0 . \end{cases}$

14. Determine the values of k for which solutions exist, and find the solutions

(i) $\begin{cases} 2x - y + 3z = 8 \\ 3x + 2y + z = 3 \\ x - 4y + 5z = k , \end{cases}$ (ii) $\begin{cases} x + 3y - 2z = k \\ 3x - y + 2z = 3 \\ 4x - 3y + 4z = 2 . \end{cases}$

15. Solve the equation

(i) $\begin{vmatrix} 3-x & 2 \\ 9 & -1-x \end{vmatrix} = 0$, (ii) $\begin{vmatrix} -x & 6 \\ 1 & 5-x \end{vmatrix} = 0$,

(iii) $\begin{vmatrix} 1 & x & 3 \\ x & -2 & 0 \\ -1 & 1 & 4 \end{vmatrix} = 0$, (iv) $\begin{vmatrix} x+2 & 3 & -1 \\ -1 & x-2 & 4 \\ 2 & 2 & x+1 \end{vmatrix} = 0$.

16. Solve the equation

(i) $\begin{vmatrix} 1-x & 0 & 0 \\ 0 & -3-x & 4 \\ 0 & 4 & 3-x \end{vmatrix} = 0$, (ii) $\begin{vmatrix} 16-x & -8 & 12 \\ -8 & 4-x & -6 \\ 12 & -6 & 9-x \end{vmatrix} = 0$.

17. In the matrix

$$\begin{pmatrix} a_{11} & a_{12} \\ a_{21} & a_{22} \end{pmatrix} ,$$

let the vectors (a_{11},a_{12}) and (a_{21},a_{22}) form an orthonormal set. Show that the determinant of this matrix is ± 1.

18.* In the matrix

$$\begin{pmatrix} a_{11} & a_{12} & a_{13} \\ a_{21} & a_{22} & a_{23} \\ a_{31} & a_{32} & a_{33} \end{pmatrix}$$

let the vectors (a_{11},a_{12},a_{13}), (a_{21},a_{22},a_{23}), and (a_{31},a_{32},a_{33}) form an orthonormal set. Show that the determinant of this matrix is ± 1.

4. SPECIAL TRANSFORMATIONS AND THEIR MATRICES

4.1 **Inverse of a linear transformation:** For a given transformation **A** there might exist a transformation **B** such that **AB** = **I**. It is customary to call this transformation the *inverse* of **A**, written $\mathbf{A}^{-1}$. Intuitively speaking $\mathbf{A}^{-1}$ is the way of going back from the vector **A**V to V.

If the matrix of **A** with respect to a base $\{U_1, U_2, U_3\}$ is

$$\begin{pmatrix} a_{11} & a_{12} & a_{13} \\ a_{21} & a_{22} & a_{23} \\ a_{31} & a_{32} & a_{33} \end{pmatrix}$$

we shall find the matrix of $\mathbf{A}^{-1}$, if it exists, as follows:

Suppose for a vector $V = (x, y, z)$ we have $\mathbf{A}V = (x_1, y_1, z_1)$, and by 2.9 (1) we have

$$\begin{cases} x_1 = a_{11}x + a_{21}y + a_{31}z \\ y_1 = a_{12}x + a_{22}y + a_{32}z \\ z_1 = a_{13}x + a_{23}y + a_{33}z \ . \end{cases}$$

Here we have to find x, y, and z in terms of x_1, y_1 and z_1. By 3.4 this system of linear equations has a unique solution, that is, the inverse exists, if and only if

$$\begin{vmatrix} a_{11} & a_{21} & a_{31} \\ a_{12} & a_{22} & a_{31} \\ a_{13} & a_{23} & a_{33} \end{vmatrix} = \begin{vmatrix} a_{11} & a_{12} & a_{13} \\ a_{21} & a_{22} & a_{23} \\ a_{31} & a_{32} & a_{33} \end{vmatrix} \neq 0 \ .$$

Solving the equations, suppose we get

$$x = b_{11}x_1 + b_{21}y_1 + b_{31}z_1 \ ,$$

$$y = b_{12}x_1 + b_{22}y_1 + b_{32}z_1 \ ,$$

$$z = b_{13}x_1 + b_{23}y_1 + b_{33}z_1 \ .$$

Writing these equations in matrix form as in 2.9 (1) we have

$$(x\ y\ z) = x_1\ y_1\ z_1) \begin{pmatrix} b_{11} & b_{12} & b_{13} \\ b_{21} & b_{22} & b_{23} \\ b_{31} & b_{32} & b_{33} \end{pmatrix} .$$

Therefore $\mathbf{A}^{-1} = \mathbf{B}$ and its matrix is

$$\begin{pmatrix} b_{11} & b_{12} & b_{13} \\ b_{21} & b_{22} & b_{23} \\ b_{31} & b_{32} & b_{33} \end{pmatrix} .$$

A transformation whose inverse exists is called *non-singular*. The reader may show that $\mathbf{A}^{-1}\mathbf{A}$ is also equal to **I**.

4.2 **A practical way of getting the inverse:** We demonstrate the method by the following example. Let the matrix of **A** be $\begin{pmatrix} 2 & 0 \\ 1 & 1 \end{pmatrix}$. The matrix of $\mathbf{A}^{-1}$ is the matrix $\begin{pmatrix} p & q \\ r & s \end{pmatrix}$ such that $\begin{pmatrix} 2 & 0 \\ 1 & 1 \end{pmatrix}\begin{pmatrix} p & q \\ r & s \end{pmatrix} = \begin{pmatrix} 1 & 0 \\ 0 & 1 \end{pmatrix}$. Multiplying, and putting the corresponding terms of the two sides equal we have

$$\begin{cases} 2p = 1 \\ 2q = 0 \\ p + r = 0 \\ q + s = 1 \end{cases}$$

Solving this system we see that the matrix of $\mathbf{A}^{-1}$ is: $\begin{pmatrix} \frac{1}{2} & 0 \\ -\frac{1}{2} & 1 \end{pmatrix}$.

4.3 **Theorem:** If **A** and **B** are two linear transformations on the space, and if $\mathbf{A}^{-1}$ and $\mathbf{B}^{-1}$ exist, then $(\mathbf{A}\,\mathbf{B})^{-1}$ exists and

$$(\mathbf{A}\,\mathbf{B})^{-1} = \mathbf{B}^{-1}\mathbf{A}^{-1} \quad \text{(note the change of order).}$$

Proof: By 2.2 we have

(1) $(\mathbf{A}\,\mathbf{B})V = \mathbf{B}(\mathbf{A}V)$.

Suppose $(\mathbf{A}\,\mathbf{B})V = U$. Then the inverse of **A B** is a transformation **C** such that

(2) $\mathbf{C}\,U = V$.

Substituting $\mathbf{C}U$ for V in (1) we get

$$(\mathbf{A}\,\mathbf{B})(\mathbf{C}\,U) = U\,, \quad \text{i.e.,}$$

(3) $\mathbf{B}\,[\mathbf{A}(\mathbf{C}U)] = U$.

Operating on both sides of (3) by $\mathbf{B}^{-1}$ we get

$$\mathbf{B}^{-1}\{\mathbf{B}[\mathbf{A}(\mathbf{C}U)]\} = \mathbf{B}^{-1}U, \quad \text{or}$$

(4) $\mathbf{A}(\mathbf{C}U) = \mathbf{B}^{-1}U$.

Now operating on both sides of (4) by $\mathbf{A}^{-1}$ we get

$$\mathbf{A}^{-1}[\mathbf{A}(\mathbf{C}U)] = \mathbf{A}^{-1}(\mathbf{B}^{-1}U) \quad \text{or}$$

(5) $\mathbf{C}U = \mathbf{A}^{-1}(\mathbf{B}^{-1}U) = \mathbf{B}^{-1}\mathbf{A}^{-1}U$.

Therefore $\mathbf{C} = \mathbf{B}^{-1}\mathbf{A}^{-1}$.

4.4 **Adjoint of a transformation:** Let **A** be a transformation on the space. A transformation **B** for which $(\mathbf{A}V, W) = (V, \mathbf{B}W)$, for all V and W, is called the *adjoint* of **A**, and we denote **B** by $\mathbf{A}'$. Note that $\mathbf{A}'' = (\mathbf{A}')' = \mathbf{A}$.

4.5 **Theorem:** Let $\begin{pmatrix} a_{11} & a_{12} & a_{13} \\ a_{21} & a_{22} & a_{23} \\ a_{31} & a_{32} & a_{33} \end{pmatrix}$ be the matrix of **A** with respect to an orthonormal base $\{U_1, U_2, U_3\}$. The matrix of $\mathbf{A}'$ with respect to the same base is

$$\begin{pmatrix} a_{11} & a_{21} & a_{31} \\ a_{12} & a_{22} & a_{32} \\ a_{13} & a_{23} & a_{33} \end{pmatrix},$$

the transpose of the matrix of **A**.

Proof: By 2.8 and 2.9 (1) we have

$$(\mathbf{A}V, W) = (x_1\ y_1\ z_1) \begin{pmatrix} a_{11} & a_{12} & a_{13} \\ a_{21} & a_{22} & a_{23} \\ a_{31} & a_{32} & a_{33} \end{pmatrix} \begin{pmatrix} x_2 \\ y_2 \\ z_2 \end{pmatrix},$$

where $V = (x_1\ y_1\ z_1)$ and $W = \begin{pmatrix} x_2 \\ y_2 \\ z_1 \end{pmatrix}$.

But by 2.9 (2) we have

$$\begin{pmatrix} a_{11} & a_{12} & a_{13} \\ a_{21} & a_{22} & a_{23} \\ a_{31} & a_{32} & a_{33} \end{pmatrix} \begin{pmatrix} x_2 \\ y_2 \\ z_2 \end{pmatrix} = \mathbf{A}'W,$$

and this proves the theorem.

4.6 **Theorem:** Let **A** and **B** be two linear transformations. Then

(1) $(\mathbf{A} + \mathbf{B})' = \mathbf{A}' + \mathbf{B}'$,

(2) $(\mathbf{A}\,\mathbf{B})' = \mathbf{B}'\mathbf{A}'$ (note the order).

Proof: We shall prove (2) and leave (1) to the reader as an exercise. By 2.2 and 4.4 we see that

$$((\mathbf{A}\,\mathbf{B})\,V, W) = (\mathbf{B}(\mathbf{A}\,V), W) = (\mathbf{A}V, \mathbf{B}'W) = (V, \mathbf{A}'(\mathbf{B}'W)) = (V, (\mathbf{B}'\mathbf{A}')\,W).$$

Therefore $(\mathbf{AB})' = \mathbf{B}'\mathbf{A}'$.

4.7 **Theorem:** Let **A** be a transformation on the space. If $\mathbf{A}^{-1}$ exists, then $(\mathbf{A}')^{-1}$ exists and $(\mathbf{A}')^{-1} = (\mathbf{A}^{-1})'$.

Proof: Clearly $\mathbf{I}' = \mathbf{I}$. This implies that $\mathbf{A}\,\mathbf{A}^{-1} = \mathbf{I} = \mathbf{I}' = (\mathbf{A}^{-1})'\,\mathbf{A}'$ [see 4.6 (2)]. Also $\mathbf{A}^{-1}\mathbf{A} = \mathbf{I} = \mathbf{I}' = \mathbf{A}'(\mathbf{A}^{-1})'$. Therefore $(\mathbf{A}^{-1})' = (\mathbf{A}')^{-1}$.

4.8 **Orthogonal (Unitary) transformations:** An *orthogonal transformation* **A** is one which preserves inner product. That is, for any two vectors U and V we have $(U, V) = (\mathbf{A}\,U, \mathbf{A}\,V)$.

If the matrix of **A** with respect to the orthonormal base $\{U_1, U_2, U_3\}$ is

$$\begin{pmatrix} a_{11} & a_{12} & a_{13} \\ a_{21} & a_{22} & a_{23} \\ a_{31} & a_{32} & a_{33} \end{pmatrix}, \text{ then}$$

I. the sum of the squares of the terms of each row is one.
II. If we consider each row as a vector, any two rows are orthogonal.

This can be proved very easily as follows: for example for U_1 and U_2 we have

$$\mathbf{A}\, U_1 = a_{11}U_1 + a_{12}U_2 + a_{13}U_3\,,$$

$$\mathbf{A}\, U_2 = a_{21}U_1 + a_{22}U_2 + a_{23}U_3\,. \quad \text{But}$$

$$1 = (U_1,U_1) = (\mathbf{A}U_1,\ \mathbf{A}U_1) = |\mathbf{A}\ U_1|^2 = a_{11}^2 + a_{12}^2 + a_{13}^2\,.$$

and

$$0 = (U_1,U_2) = (\mathbf{A}\ U_1,\ \mathbf{A}\ U_2) = a_{11}\ a_{21} + a_{12}\ a_{22} + a_{13}\ a_{23}\,.$$

4.9 **Theorem:** If $\begin{pmatrix} a_{11} & a_{12} & a_{13} \\ a_{21} & a_{22} & a_{23} \\ a_{31} & a_{32} & a_{33} \end{pmatrix}$ is the matrix of a transformation $\mathbf{A}$ with respect to an orthonormal base, then $\mathbf{A}' = \mathbf{A}^{-1}$ if and only if the rows of the matrix form an orthcnormal set of vectors.

Proof: The matrix of $\mathbf{A}'$ is

$$\begin{pmatrix} a_{11} & a_{21} & a_{31} \\ a_{12} & a_{22} & a_{32} \\ a_{13} & a_{23} & a_{33} \end{pmatrix}.$$

Consequently, if the rows of the matrix of $\mathbf{A}$ are an orthonormal set of vectors, the matrix of $\mathbf{AA}'$ is $\begin{pmatrix} 1 & 0 & 0 \\ 0 & 1 & 0 \\ 0 & 0 & 1 \end{pmatrix}$, so that $\mathbf{A}\ \mathbf{A}' = \mathbf{I}$ and $\mathbf{A}' = \mathbf{A}^{-1}$. Conversely, if $\mathbf{A}\ \mathbf{A}' = \mathbf{I}$, then

$$\begin{pmatrix} a_{11} & a_{12} & a_{13} \\ a_{21} & a_{22} & a_{23} \\ a_{31} & a_{32} & a_{33} \end{pmatrix} \begin{pmatrix} a_{11} & a_{21} & a_{31} \\ a_{12} & a_{22} & a_{32} \\ a_{13} & a_{23} & a_{33} \end{pmatrix} = \begin{pmatrix} 1 & 0 & 0 \\ 0 & 1 & 0 \\ 0 & 0 & 1 \end{pmatrix},$$

and, since the columns of the matrix of $\mathbf{A}'$ are the rows of the matrix of $\mathbf{A}$, the row vectors of the matrix of $\mathbf{A}$ are orthonormal set. Furthermore it is now obvious (since $\mathbf{A}'\mathbf{A} = \mathbf{I}$) that the column vectors also form an orthonormal set.

Note that for a unitary transformation the inverse always exists. We see also that if $\mathbf{A}' = \mathbf{A}^{-1}$ then the transformation $\mathbf{A}$ is unitary. For

$$(\mathbf{A}U,\ \mathbf{A}V) = (U,\mathbf{A}'(\mathbf{A}V)) = (U,\ \mathbf{A}^{-1}\ (\mathbf{A}V)) = (U,(\mathbf{AA}^{-1})\,V) = (U,\ V)\,.$$

4.10 **Change of Base:** Let $\{U_1,\ U_2,\ U_3\}$ be a base of the space, not necessarily orthonormal. Let $\{V_1,\ V_2,\ V_3\}$ be another base. Let $\mathbf{A}$ be a linear transformation on the space.

Let $[\mathbf{A}] = \begin{pmatrix} a_{11} & a_{12} & a_{13} \\ a_{21} & a_{22} & a_{23} \\ a_{31} & a_{32} & a_{33} \end{pmatrix}$ be the matrix of $\mathbf{A}$ with respect to $\{U_1, U_2, U_3\}$ and $[\mathbf{B}] = \begin{pmatrix} b_{11} & b_{12} & b_{13} \\ b_{21} & b_{22} & b_{23} \\ b_{31} & b_{32} & b_{33} \end{pmatrix}$

be the matrix of **A** with respect to $\{V_1, V_2, V_3\}$. Then we would like to know the relation of the two matrices.

Suppose

$$\begin{aligned} V_1 &= c_{11}\,U_1 + c_{12}\,U_2 + c_{13}\,U_3\,, \\ (1)\qquad V_2 &= c_{21}\,U_1 + c_{22}\,U_2 + c_{23}\,U_3\,, \\ V_3 &= c_{31}\,U_1 + c_{32}\,U_2 + c_{33}\,U_3\,. \end{aligned}$$

It is observed that there is a matrix $[\mathbf{C}] = \begin{pmatrix} c_{11} & c_{12} & c_{13} \\ c_{21} & c_{22} & c_{23} \\ c_{31} & c_{32} & c_{33} \end{pmatrix}$ corresponding to (1).

Let X be a vector, with $X = xU_1 + yU_2 + zU_3$ and also $X = \xi V_1 + \eta V_2 + \zeta V_3$.

$$\begin{aligned} \xi V_1 + \eta V_2 + \zeta V_3 &= \xi\,(c_{11}U_1 + c_{12}U_2 + c_{13}U_3) + \eta\,(c_{21}U_1 + c_{22}U_2 + c_{23}U_3) + \zeta\,(c_{31}U_1 + c_{32}U_2 + c_{33}U_3) \\ &= (c_{11}\xi + c_{21}\eta + c_{31}\zeta)\,U_1 + (c_{12}\xi + c_{22}\eta + c_{32}\zeta)\,U_2 + (c_{13}\xi + c_{23}\eta + c_{33}\zeta)\,U_3 \\ &= xU_1 + yU_2 + zU_3\,. \end{aligned}$$

We conclude that

$$(x\ y\ z) = (\xi\ \eta\ \zeta)\begin{pmatrix} c_{11} & c_{12} & c_{13} \\ c_{21} & c_{22} & c_{23} \\ c_{31} & c_{32} & c_{33} \end{pmatrix} \quad \text{or} \quad (x\ y\ z) = (\xi\ \eta\ \zeta)\,[\mathbf{C}]\,.$$

From (1) we can show that $[\mathbf{C}]^{-1}$ exists. Therefore

$$(\xi\ \eta\ \zeta) = (x\ y\ z)\,[\mathbf{C}]^{-1}\,.$$

Let us write $\mathbf{A}\,X = Y = x'U_1 + y'U_2 + z'U_3 = \xi' V_1 + \eta' V_2 + \zeta' V_3$. Then

$$(\xi'\ \eta'\ \zeta') = (x'\ y'\ z')\,[\mathbf{C}^{-1}] = (x\ y\ z)\,[\mathbf{A}][\mathbf{C}^{-1}] = (\xi\ \eta\ \zeta)[\mathbf{C}][\mathbf{A}][\mathbf{C}^{-1}]\,.$$

Hence $\quad [\mathbf{B}] = [\mathbf{C}][\mathbf{A}][\mathbf{C}^{-1}]\,.$

Note that the operation of changing base described in (1) is a linear transformation with matrix $[\mathbf{C}]$.

4.11 **Theorem:** The change from one orthonormal base to another is unitary.

Proof: By 1.13, since $\{V_1, V_2, V_3\}$ is orthonormal, from 4.10 (1) we have $1 = (V_i, V_i) = c_{i1}^{\,2} + c_{i2}^{\,2} + c_{i3}^{\,2}$, while

$$0 = (V_i, V_j) = c_{i1}c_{j1} + c_{i2}c_{j2} + c_{i3}c_{j3}, \quad \text{if } i \neq j,\ i,j = 1,2,3\,.$$

Consequently by 4.9 the transformation **C** is unitary.

Illustration 1: The matrix of a transformation **A** with respect to the base $\{(1,0), (0,1)\}$ is

$$\begin{pmatrix} 1 & 3 \\ 0 & -2 \end{pmatrix}.$$

Find the matrix of **A** with respect to the base $\{(2,1),(1,-1)\}$.

We see that

$$[\mathbf{C}] = \begin{pmatrix} 2 & 1 \\ 1 & -1 \end{pmatrix}.$$

Let

$$[\mathbf{C}^{-1}] = \begin{pmatrix} p & q \\ r & s \end{pmatrix},$$

so that

$$\begin{pmatrix} p & q \\ r & s \end{pmatrix} \begin{pmatrix} 2 & 1 \\ 1 & -1 \end{pmatrix} = \begin{pmatrix} 1 & 0 \\ 0 & 1 \end{pmatrix}.$$

Then

$$\begin{cases} 2p + q = 1 \\ p - q = 0, \end{cases} \qquad \begin{cases} 2r + s = 0 \\ r - s = 1, \end{cases}$$

and $p = 1/3$, $q = 1/3$, $r = 1/3$, and $s = -2/3$.

Let the matrix of **A** with respect to the base $\{(2,1),(1,-1)\}$ by $[\mathbf{B}]$. Then

$$[\mathbf{B}] = [\mathbf{C}][\mathbf{A}][\mathbf{C}^{-1}] = \begin{pmatrix} 2 & 1 \\ 1 & -1 \end{pmatrix} \begin{pmatrix} 1 & 3 \\ 0 & -2 \end{pmatrix} \begin{pmatrix} \frac{1}{3} & \frac{1}{3} \\ \frac{1}{3} & -\frac{2}{3} \end{pmatrix} = \begin{pmatrix} 2 & -2 \\ 2 & -3 \end{pmatrix}.$$

Illustration 2: The matrix of a transformation **A** with respect to the orthonormal base $\{(1,0,0),(0,1,0),(0,0,1)\}$ is

$$\begin{pmatrix} 1 & -1 & 0 \\ 2 & 1 & 1 \\ -2 & 0 & 1 \end{pmatrix}.$$

Find the matrix of **A** with respect to the base $\left\{\left(-\frac{2}{3},\frac{1}{3},\frac{2}{3}\right),\left(-\frac{1}{3},\frac{2}{3},-\frac{2}{3}\right),\left(\frac{2}{3},\frac{2}{3},\frac{1}{3}\right)\right\}$.

We observe that

$$[\mathbf{C}] = \begin{pmatrix} -\frac{2}{3} & \frac{1}{3} & \frac{2}{3} \\ -\frac{1}{3} & \frac{2}{3} & -\frac{2}{3} \\ \frac{2}{3} & \frac{2}{3} & \frac{1}{3} \end{pmatrix}.$$

Since both bases are orthonormal, $[\mathbf{C}^{-1}] = [\mathbf{C}]'$, hence the matrix $[\mathbf{B}]$ of the transformation **A** with respect to the new base can be found as follows:

$$[\mathbf{B}] = [\mathbf{C}][\mathbf{A}][\mathbf{C}]' =$$

$$\begin{pmatrix} -\frac{2}{3} & \frac{1}{3} & \frac{2}{3} \\ -\frac{1}{3} & \frac{2}{3} & -\frac{2}{3} \\ \frac{2}{3} & \frac{2}{3} & \frac{1}{3} \end{pmatrix} \begin{pmatrix} 1 & -1 & 0 \\ 2 & 1 & 1 \\ -2 & 0 & 1 \end{pmatrix} \begin{pmatrix} -\frac{2}{3} & -\frac{1}{3} & \frac{2}{3} \\ \frac{1}{3} & \frac{2}{3} & \frac{2}{3} \\ \frac{2}{3} & -\frac{2}{3} & \frac{1}{3} \end{pmatrix} = \begin{pmatrix} \frac{17}{9} & \frac{4}{9} & \frac{1}{9} \\ -\frac{11}{9} & -\frac{1}{9} & \frac{20}{9} \\ -\frac{2}{9} & -\frac{10}{9} & \frac{11}{9} \end{pmatrix}.$$

EXERCISES 4

1. Prove that if the transformation **A** is non-singular, then the matrix of $\mathbf{A}^{-1}$ is

$$\frac{1}{\det.\mathbf{A}}\begin{pmatrix} \mathbf{A}_{11} & \mathbf{A}_{21} & \mathbf{A}_{31} \\ \mathbf{A}_{12} & \mathbf{A}_{22} & \mathbf{A}_{32} \\ \mathbf{A}_{13} & \mathbf{A}_{23} & \mathbf{A}_{33} \end{pmatrix},$$

where $\mathbf{A}_{ij}$ is the cofactor of a_{ij} of the matrix **A**.

2. Verify 4.3 by matrices.
3. Verify 4.6 by matrices.
4. Show that for a unitary transformation **A**, det **A** $= \pm 1$, and the cofactor of any term a_{ij} is $\pm\, a_{ij}$.
5. Show that the following transformations are unitary.

   (i) rotation about the origin in the plane.
   (ii) rotation about any axis through the origin in the space.
   (iii) symmetry with respect to any axis through the origin.
   (iv) symmetry with respect to any plane through the origin.

6. Find the inverse of:

(i) $\begin{pmatrix} 1 & 2 \\ 3 & 4 \end{pmatrix}$, (ii) $\begin{pmatrix} 0 & 1 & 2 \\ -1 & 0 & 1 \\ 5 & 7 & 8 \end{pmatrix}$, (iii) $\begin{pmatrix} a & b \\ c & d \end{pmatrix}$ if $ad - bc \neq 0$.

7. Find the inverse of

(i) $\begin{pmatrix} \sin\theta & \cos\theta \\ -\cos\theta & \sin\theta \end{pmatrix}$, (ii) $\begin{pmatrix} 1 & 0 & 0 \\ 0 & \sin\theta & \cos\theta \\ 0 & -\cos\theta & \sin\theta \end{pmatrix}$, (iii) $\begin{pmatrix} \sin\theta\cos\phi & \sin\theta\sin\phi & \cos\theta \\ \cos\theta\cos\phi & \cos\theta\sin\phi & -\sin\theta \\ \sin\phi & -\cos\phi & 0 \end{pmatrix}$.

8. The matrix of **A** with respect to the base $\{(1,0), (0,1)\}$ is

$$\begin{pmatrix} -2 & 5 \\ 7 & 0 \end{pmatrix}.$$

Find the matrix of **A** with respect to the base $\{(1,1), (-1,0)\}$.

9. The matrix of **A** with respect to $\{(1,0,0), (0,1,0), (0,0,1)\}$ is

$$\begin{pmatrix} -1 & 0 & 2 \\ 7 & -1 & -1 \\ 2 & 3 & 4 \end{pmatrix}.$$

Find the matrix of **A** with respect to $\{(1,1,1), (0,0,-1), (2,-1,0)\}$.

10. If **A** and **B** are two unitary transformations, show that **AB** and **BA** are also unitary.
11. Show that the adjoint of a projection on the plane $ax + by + cz = 0$ is itself.
12. Show that the adjoint of a projection on a line through the origin is itself.
13. Let the matrix of **A** with respect to the orthonormal base $\{(1,0), (0,1)\}$ be

$$\begin{pmatrix} 2 & 2 \\ 2 & 5 \end{pmatrix}.$$

If the coordinate system is rotated through an acute angle θ with $\sin\theta = \frac{2\sqrt{5}}{5}$, find the matrix of **A** with respect to the new base.

14. With respect to the base $\{(1,0), (0,1)\}$, $V = (3,-1)$. Find the components of V if the coordinate system is changed as in exercise 13.
15. Let the equations of the transformation **A** be

$$\begin{cases} x' = 3x + 2y - z \\ y' = 2x - y \\ z' = x + z . \end{cases}$$

Find the equations of the adjoint of **A**.
16. Show that the symmetry with respect to any subspace is its own inverse. Give both geometric and matrix proofs.
17. The matrix of a transformation **A** with respect to the base $\{(1,0,0), (0,1,0), (0,0,1)\}$ is

$$\begin{pmatrix} 3 & 0 & 1 \\ 0 & 2 & 0 \\ 1 & 0 & 3 \end{pmatrix} .$$

Find a base with respect to which the matrix of **A** becomes

(i) $\begin{pmatrix} 2 & 0 & 0 \\ 0 & 2 & 0 \\ 0 & 0 & 4 \end{pmatrix}$ (ii) $\begin{pmatrix} 2 & 0 & 0 \\ 0 & 4 & 0 \\ 0 & 0 & 2 \end{pmatrix}$ (iii) $\begin{pmatrix} 4 & 0 & 0 \\ 0 & 2 & 0 \\ 0 & 0 & 2 \end{pmatrix}$.

18. Supply the missing elements so that the matrix will be unitary

(i) $\begin{pmatrix} \frac{\sqrt{2}}{2} & \frac{\sqrt{2}}{2} \\ \cdot & \cdot \end{pmatrix}$ (ii) $\begin{pmatrix} \cdot & \cdot \\ \frac{-1}{2} & \frac{\sqrt{3}}{2} \end{pmatrix}$ (iii) $\begin{pmatrix} \sin\alpha & \cdot \\ \cos\alpha & \cdot \end{pmatrix}$.

19. Supply the missing elements so that the matrix will be unitary

(i) $\begin{pmatrix} \cdot & \cdot & \frac{2}{3} \\ \frac{-1}{3} & \frac{2}{3} & \cdot \\ \frac{2}{3} & \cdot & \frac{1}{3} \end{pmatrix}$, (ii) $\begin{pmatrix} \frac{3}{7} & \cdot & \frac{-6}{7} \\ \cdot & \cdot & \frac{-2}{7} \\ \frac{2}{7} & \frac{6}{7} & \cdot \end{pmatrix}$.

ADDITIONAL PROBLEMS 4

1. Find the inverse if it exists:

(i) $\begin{pmatrix} 0 & 1 \\ 2 & 2 \end{pmatrix}$, (ii) $\begin{pmatrix} \sin\theta & \cos\theta \\ \cos\theta & \sin\theta \end{pmatrix}$.

2. Find the inverse if it exists:

(i) $\begin{pmatrix} 2 & 1 & 1 \\ -2 & -3 & 0 \\ 0 & 0 & 3 \end{pmatrix}$, (ii) $\begin{pmatrix} -1 & -1 & -1 \\ 0 & 1 & 0 \\ 2 & 3 & 5 \end{pmatrix}$.

3. Let the base be changed from {(1,0), (0,1)} to {(2,2), (0,–1)}. Find the matrix of **A** with respect to the second base if the matrix of **A** with respect to the first base is:

(i) $\begin{pmatrix} 3 & 0 \\ 1 & 2 \end{pmatrix}$, (ii) $\begin{pmatrix} -1 & -1 \\ 7 & 0 \end{pmatrix}$, (iii) $\begin{pmatrix} -2 & -7 \\ -1 & -3 \end{pmatrix}$, (iv) $\begin{pmatrix} a & 2a \\ 3a & 4a \end{pmatrix}$.

4. Let two bases for the space be {(1,0,0), (0,1,0), (0,0,1)} and {(0,1,1), (–1,–1,0), (1,0,1)}. Find the matrix of **A** with respect to the second base if the matrix of **A** with respect to the first base is:

(i) $\begin{pmatrix} 2 & 1 & 0 \\ -1 & 0 & 2 \\ 7 & 1 & 1 \end{pmatrix}$, (ii) $\begin{pmatrix} -2 & -2 & -2 \\ -1 & 0 & -1 \\ 5 & 5 & 1 \end{pmatrix}$.

5. Consider the two bases of problem 3. Find the matrix of **A** with respect to the first base if the matrix of **A** with respect to the second base is:

(i) $\begin{pmatrix} 1 & -2 \\ 0 & 1 \end{pmatrix}$, (ii) $\begin{pmatrix} 7 & 7 \\ -2 & -1 \end{pmatrix}$.

6. Consider the two bases of problem 4. Find the matrix of **A** with respect to the first base if the matrix of **A** with respect to the second base is:

(i) $\begin{pmatrix} 0 & 1 & -1 \\ -1 & 2 & 7 \\ 3 & 0 & 0 \end{pmatrix}$, (ii) $\begin{pmatrix} 3 & 1 & 0 \\ 7 & 6 & 0 \\ 1 & 1 & 7 \end{pmatrix}$.

7. Supply the missing elements in order to get a unitary matrix

(i) $\begin{pmatrix} \frac{1}{3} & \cdot & \frac{2}{3} \\ \cdot & \frac{1}{3} & \frac{2}{3} \\ \cdot & \cdot & \cdot \end{pmatrix}$, (ii) $\begin{pmatrix} \frac{1}{\sqrt{2}} & \frac{-1}{\sqrt{2}} & \cdot \\ \frac{1}{\sqrt{3}} & \cdot & \cdot \\ \cdot & \cdot & \cdot \end{pmatrix}$,

(iii) $\begin{pmatrix} \frac{1}{\sqrt{6}} & \frac{-1}{\sqrt{6}} & \cdot \\ \frac{1}{\sqrt{2}} & \cdot & 0 \\ \cdot & \cdot & \cdot \end{pmatrix}$, (iv) $\begin{pmatrix} \frac{6}{19} & \cdot & \frac{15}{19} \\ \frac{10}{19} & \cdot & \frac{6}{19} \\ \cdot & \cdot & \cdot \end{pmatrix}$.

8. Let the matrix of **A** with respect to {(1,0), (0,1)} be

$$\begin{pmatrix} 1 & 0 \\ 0 & 6 \end{pmatrix}.$$

Find a base with respect to which the matrix of **A** becomes

$$\begin{pmatrix} 2 & 2 \\ 2 & 5 \end{pmatrix}.$$

9. Let the matrix of **A** with respect to $\{(1,0,0), (0,1,0), (0,0,1)\}$ be

$$\begin{pmatrix} 2 & 0 & 0 \\ 0 & 2 & 0 \\ 0 & 0 & 4 \end{pmatrix}.$$

Find the base with respect to which the matrix of **A** is

$$\begin{pmatrix} 3 & 0 & 1 \\ 0 & 2 & 0 \\ 1 & 0 & 3 \end{pmatrix}.$$

10.* Show that if a transformation **A** preserves norm then it preserves inner product.

11. Which of the following transformations in the plane

(a) are unitary? (b) preserve angle?

(i) rotation through an angle θ ,
(ii) homothetic transformation, i.e., $\mathbf{A}U = kU$, k a fixed scalar ,
(iii) projection on a subspace ,
(iv) symmetry with respect to the line $ax + by = 0$.

12. Which of the following transformations in the space

(a) are unitary? (b) preserve angle?

(i) rotation about the x-axis through an angle θ ,
(ii) homothetic transformation ,
(iii) projection on a subspace ,
(iv) symmetry with respect to a subspace .

5. CHARACTERISTIC EQUATION OF A TRANSFORMATION AND QUADRATIC FORMS

5.1 **Characteristic values and characteristic vectors of a transformation:** Let **A** be a linear transformation on the space. We ask the following question:

Is there any vector V in the space such that for some scalar m, $\mathbf{A}V = mV$? The answer is given by the following.

Let the matrix of **A** with respect to a base $\{U_1, U_2, U_3\}$ be

$$\begin{pmatrix} a_{11} & a_{12} & a_{13} \\ a_{21} & a_{22} & a_{23} \\ a_{31} & a_{32} & a_{33} \end{pmatrix} .$$

Then the equation

(1) $\mathbf{A}V = mV$ will be

(2) $$(x\ y\ z) \begin{pmatrix} a_{11} & a_{12} & a_{13} \\ a_{21} & a_{22} & a_{23} \\ a_{31} & a_{32} & a_{33} \end{pmatrix} = (mx\ \ my\ \ mz) .$$

This yields the following system of linear equations:

(3) $$\begin{cases} (a_{11} - m)x + a_{21}y + a_{31}z = 0 \\ a_{12}x + (a_{22} - m)y + a_{32}z = 0 \\ a_{13}x + a_{23}y + (a_{33} - m)z = 0 \end{cases} .$$

In order to have non-zero solutions for (3) by 3.4, we have to have

(4) $$\begin{vmatrix} a_{11} - m & a_{12} & a_{13} \\ a_{21} & a_{22} - m & a_{23} \\ a_{31} & a_{32} & a_{33} - m \end{vmatrix} = 0 .$$

It is customary to write (4) as $|\mathbf{A} - m\mathbf{I}| = 0$. Clearly (4) is an equation of the third degree in m called the *characteristic equation* of **A**. Let m_1, m_2, and m_3 be the roots of (4), real or complex, single or multiple. We call m_1, m_2, and m_3 the *characteristic values* of **A**. Other names such as *Eigenvalues, proper values,* etc., have also been used.

A vector V for which $\mathbf{A}V = m_iV$ is called a *characteristic vector* corresponding to m_i, $i = 1,2,3$.

5.2 **Theorem:** The characteristic values and the characteristic equation of a transformation are independent of the choice of base.

Proof: Consider $\mathbf{A}V = mV$. Clearly this can be written as $(\mathbf{A} - m\mathbf{I})V = 0$. Suppose **C** is a change of base. Let the matrix of $\mathbf{A} - m\mathbf{I}$ with respect to the new base be $[\mathbf{B}]$, and the matrix of **C** be $[\mathbf{C}]$. Let $[\mathbf{A} - m\mathbf{I}]$ be the matrix of $\mathbf{A} - m\mathbf{I}$ with respect to the old base. By 4.10 and 2.3 we have

(1) $[\mathbf{CAC}^{-1} - m\mathbf{I}] = [\mathbf{CAC}^{-1} - m\mathbf{CIC}^{-1}] = [\mathbf{C}][\mathbf{A} - m\mathbf{I}][\mathbf{C}^{-1}] = [\mathbf{B}]$.

But by 3.3 we have

$$\det[\mathbf{B}] = \det[\mathbf{C}]\det[\mathbf{A} - m\mathbf{I}]\det[\mathbf{C}^{-1}] = \det[\mathbf{C}]\det[\mathbf{C}^{-1}]\det[\mathbf{A} - m\mathbf{I}]$$
$$= \det[\mathbf{C}\mathbf{C}^{-1}]\det[\mathbf{A} - m\mathbf{I}] = \det[\mathbf{A} - m\mathbf{I}] \ ,$$

where for example $\det[\mathbf{A}]$ means the determinant of **A**.

This shows that the characteristic equation is independent of the choice of base. Therefore the characteristic values are also independent of the choice of base.

5.3 **Definition:** The *trace* of a matrix is the sum of the elements in the main diagonal. For example for

$$[\mathbf{A}] = \begin{pmatrix} a_{11} & a_{12} & a_{13} \\ a_{21} & a_{22} & a_{23} \\ a_{31} & a_{32} & a_{33} \end{pmatrix}$$

we have trace $[\mathbf{A}] = a_{11} + a_{22} + a_{33}$.

5.4 **Theorem:** The trace of a matrix is equal to the sum of its characteristic values and the determinant of a matrix is equal to the product of its characteristic values.

Proof: Let us recall the following concerning the relations between roots and coefficients of a third degree equation. If x_1, x_2 and x_3 are the three roots of $x^3 + ax^2 + bx + c = 0$, then

$$x_1 + x_2 + x_3 = -a \ ,$$

$$x_1 x_2 x_3 = -c \ .$$

Now to prove the theorem we shall expand (4) of 5.1 and we get

$$m^3 - (a_{11} + a_{22} + a_{33})\, m^2 + (a_{11}a_{22} + a_{22}a_{33} + a_{33}a_{11} - a_{12}a_{21} - a_{13}a_{31} - a_{23}a_{32})\, m$$
$$- \begin{vmatrix} a_{11} & a_{12} & a_{13} \\ a_{21} & a_{22} & a_{23} \\ a_{31} & a_{32} & a_{33} \end{vmatrix} = 0$$

which proves the theorem.

5.5 **Theorem:** If V_1 and V_2 are characteristic vectors of **A** corresponding to the characteristic value m, then $V = hV_1 + kV_2$, h and k scalars, is also a characteristic vector of **A** corresponding to m.

Proof: By 2.1

$$\mathbf{A}V = h\mathbf{A}V_1 + k\mathbf{A}V_2 = hmV_1 + kmV_2 = m(hV_1 + kV_2) \ .$$

This theorem enables us to choose characteristic vectors of unit length corresponding to a characteristic value m, namely

$$V = \frac{1}{|V_1|} V_1 \ .$$

5.6 **Special transformations:**

(1) A transformation **A** is called *normal* if $\mathbf{A}\mathbf{A}' = \mathbf{A}'\mathbf{A}$.

(2) A transformation **A** is called *selfadjoint (Hermitian)* if $\mathbf{A}' = \mathbf{A}$. Note that the matrix of a selfadjoint transformation with respect to any orthonormal base is *symmetric*, i.e., it is equal to its transpose, [see 4.4]. It is clear that any selfadjoint transformation is normal.

5.7 **Change of a matrix to diagonal form:** We would like to change the base so that the matrix of a transformation **A** will have zeros everywhere except on the main diagonal. This is called *diagonal form*. For convenience suppose the matrix of each transformation is denoted by itself. By 4.10 the matrix of the transformation **A** after the change of base is given by

(1) $\mathbf{D}^{-1}\mathbf{A}\mathbf{D} = \Lambda$, where $\mathbf{D} = \mathbf{C}^{-1}$, and Λ is a matrix in diagonal form. Multiplying both sides of (1) on the right by **D** we get

(2) $\mathbf{A}\mathbf{D} = \mathbf{D}\Lambda$; or in matrix form,

$$(3)\qquad \begin{pmatrix} a_{11} & a_{12} & a_{13} \\ a_{21} & a_{22} & a_{23} \\ a_{31} & a_{32} & a_{33} \end{pmatrix}\begin{pmatrix} d_{11} & d_{12} & d_{13} \\ d_{21} & d_{22} & d_{23} \\ d_{31} & d_{32} & d_{33} \end{pmatrix} = \begin{pmatrix} d_{11} & d_{12} & d_{13} \\ d_{21} & d_{22} & d_{23} \\ d_{31} & d_{32} & d_{33} \end{pmatrix}\begin{pmatrix} \lambda_1 & 0 & 0 \\ 0 & \lambda_2 & 0 \\ 0 & 0 & \lambda_3 \end{pmatrix}.$$

We get

$$(4)\qquad \begin{pmatrix} a_{11}d_{11} + a_{12}d_{21} + a_{13}d_{31} & a_{11}d_{12} + a_{12}d_{22} + a_{13}d_{32} & a_{11}d_{13} + a_{12}d_{23} + a_{13}d_{33} \\ a_{21}d_{11} + a_{22}d_{21} + a_{23}d_{31} & a_{21}d_{12} + a_{22}d_{22} + a_{23}d_{32} & a_{21}d_{13} + a_{22}d_{23} + a_{23}d_{33} \\ a_{31}d_{11} + a_{32}d_{21} + a_{33}d_{31} & a_{31}d_{12} + a_{32}d_{22} + a_{33}d_{32} & a_{31}d_{13} + a_{32}d_{23} + a_{33}d_{33} \end{pmatrix}$$

$$= \begin{pmatrix} d_{11}\lambda_1 & d_{12}\lambda_2 & d_{13}\lambda_3 \\ d_{21}\lambda_1 & d_{22}\lambda_2 & d_{23}\lambda_3 \\ d_{31}\lambda_1 & d_{32}\lambda_2 & d_{33}\lambda_3 \end{pmatrix}.$$

Equating corresponding elements of the two sides of (4) we get the following three systems of linear homogeneous equations

$$(5)\qquad \begin{cases} (a_{11} - \lambda_1)d_{11} + a_{12}d_{21} + a_{13}d_{31} = 0 \\ a_{21}d_{11} + (a_{22} - \lambda_1)d_{21} + a_{23}d_{31} = 0 \\ a_{31}d_{11} + a_{32}d_{21} + (a_{33} - \lambda_1)d_{31} = 0 , \end{cases}$$

$$(6)\qquad \begin{cases} (a_{11} - \lambda_2)d_{12} + a_{12}d_{22} + a_{13}d_{32} = 0 \\ a_{21}d_{12} + (a_{22} - \lambda_2)d_{22} + a_{23}d_{32} = 0 \\ a_{31}d_{12} + a_{32}d_{22} + (a_{33} - \lambda_2)d_{32} = 0 , \end{cases}$$

$$(7)\qquad \begin{cases} (a_{11} - \lambda_3)d_{13} + a_{12}d_{23} + a_{13}d_{33} = 0 \\ a_{21}d_{13} + (a_{22} - \lambda_3)d_{23} + a_{23}d_{33} = 0 \\ a_{31}d_{13} + a_{32}d_{23} + (a_{33} - \lambda_3)d_{33} = 0 . \end{cases}$$

By 3.4 non-zero solutions of (5), (6) and (7) exist if $\lambda_1, \lambda_2, \lambda_3$ are solutions of

$$(8)\qquad \begin{vmatrix} a_{11} - \lambda & a_{12} & a_{13} \\ a_{21} & a_{22} - \lambda & a_{23} \\ a_{31} & a_{32} & a_{33} - \lambda \end{vmatrix} = 0 .$$

That is, λ_1, λ_2, and λ_3 are the characteristic values of **A**. Using these values we can solve (5), (6) and (7) and find **D**. If a **D** thus found is non-singular it leads to a new base in which the transformation has a diagonal matrix.

5.8 **Theorem:** For a selfadjoint transformation the characteristic values are all real, and the characteristic vectors corresponding to two distinct characteristic values are orthogonal.

Proof: Let **A** be a selfadjoint transformation. Then by 5.6 the matrix of **A** with respect to some orthonormal base is symmetric. Let

$$[\mathbf{A}] = \begin{pmatrix} a_{11} & p & q \\ p & a_{22} & r \\ q & r & a_{33} \end{pmatrix} .$$

Then (3) of 5.1 will be

$$(1) \qquad \begin{cases} a_{11}x + py + qz = mx \\ px + a_{22}y + rz = my \\ qx + ry + a_{33} = mz \end{cases} .$$

Let $m = a + ib$ be a characteristic value of **A**, and $(x_1 + ix_2, y_1 + iy_2, z_1 + iz_2)$ be a non-zero solution of (1) corresponding to m.

Substituting these values in (1) we get

$$a_{11}(x_1 + ix_2) + p(y_1 + iy_2) + q(z_1 + iz_2) = (a + ib)(x_1 + ix_2)$$

$$p(x_1 + ix_2) + a_{22}(y_1 + iy_2) + r(z_1 + iz_2) = (a + ib)(y_1 + iy_2)$$

$$q(x_1 + ix_2) + r(y_1 + iy_2) + a_{33}(z_1 + iz_2) = (a + ib)(z_1 + iz_2) .$$

Equating the real and imaginary parts we get

$$(2) \qquad \begin{cases} a_{11}x_1 + py_1 + qz_1 = ax_1 - bx_2 \\ p\,x_1 + a_{22}y_1 + rz_1 = ay_1 - by_2 \\ qx_1 + ry_1 + a_{33}z_1 = az_1 - bz_2 \\ a_{11}x_2 + py_2 + qz_2 = ax_2 + bx_1 \\ px_2 + a_{22}y_2 + rz_2 = ay_2 + by_1 \\ qx_2 + ry_2 + a_{33}z_2 = az_2 + bz_1 \end{cases} .$$

Multiplying the equations of (2) respectively by $-x_2$, $-y_2$, $-z_2$, x_1, y_1 and z_1 and adding we get $0 = b(x_2^2 + y_2^2 + z_2^2 + x_1^2 + y_1^2 + z_1^2)$. Since x_1, y_1, z_1, x_2, y_2 and z_2 are real numbers and at least one of them is is different from zero, we have $b = 0$. This proves the first part of the theorem.

Now let $m_1 \neq m_2$ be two characteristic values of **A**. Let V_1 and V_2 be two non-zero characteristic vectors of **A** corresponding to m_1 and m_2 respectively. Then $\mathbf{A}V_1 = m_1V_1$ and $\mathbf{A}V_2 = m_2V_2$.

Therefore

$$(\mathbf{A}V_1, V_2) = (m_1V_1, V_2) = m_1(V_1, V_2) ,$$

and by 4.4, since $\mathbf{A}' = \mathbf{A}$

$$(\mathbf{A}V_1,V_2) = (V_1,\mathbf{A}'V_2) = (V_1,\mathbf{A}V_2) = (V_1,m_2V_2) = m_2(V_1,V_2)\ .$$

Thus $$m_1(V_1,V_2) - m_2(V_1,V_2) = (m_1 - m_2)(V_1,V_2) = 0\ .$$

Since $$m_1 - m_2 \neq 0, \text{ we have } (V_1,V_2) = 0\ ,$$

that is, V_1 and V_2 are orthogonal.

5.9 **Definition:** A selfadjoint transformation with positive (non-negative) characteristic values is called *positive (non-negative).*

5.10 **Theorem:** For any symmetric matrix [**A**], a change to diagonal form is always possible. Moreover the change of base is a unitary transformation.

Proof: We prove the theorem for three cases.

(1) Let λ_1, λ_2, λ_3, the characteristic values of **A**, be distinct. Let V_1, V_2, and V_3 be characteristic vectors of unit length corresponding to λ_1, λ_2, λ_3 respectively. By 5.8 $\{V_1, V_2, V_3\}$ is an orthonormal set. Consider the change of base to $\{V_1, V_2, V_3\}$. Suppose [**P**] is the matrix of that change of base. By 4.11, **P** is unitary and therefore $[\mathbf{P}^{-1}] = [\mathbf{P}']$. Let $[\mathbf{P}]\,[\mathbf{A}]\,[\mathbf{P}'] = [\mathbf{B}]$ be the matrix of **A** with respect to $\{V_1, V_2, V_3\}$. Since $[\mathbf{B}'] = [\mathbf{P}]\,[\mathbf{A}']\,[\mathbf{P}']$, by 4.6, and $[\mathbf{A}'] = [\mathbf{A}]$, [**B**] is also symmetric. With respect to $\{V_1, V_2, V_3\}$ we have $V_1 = (1,0,0)$, $V_2 = (0,1,0)$, and $V_3 = (0,0,1)$.

Thus $$\mathbf{B}V_1 = \lambda_1 V_1 = (\lambda_1,0,0)\ ,$$

$$\mathbf{B}V_2 = \lambda_2 V_2 = (0,\lambda_2,0)\ ,$$

$$\mathbf{B}V_3 = \lambda_3 V_3 = (0,0,\lambda_3)\ , \quad \text{and we have}$$

$$[\mathbf{B}] = \begin{pmatrix} \lambda_1 & 0 & 0 \\ 0 & \lambda_2 & 0 \\ 0 & 0 & \lambda_3 \end{pmatrix}.$$

(2) Suppose $\lambda_1 \neq \lambda_2 = \lambda_3$ are the characteristic values of **A**. Let V_1 and V_2 be characteristic vectors of unit length corresponding to λ_1 and λ_2 respectively. Choose V_3 so that $\{V_1, V_2, V_3\}$ is an orthonormal set. Let [**P**] be the matrix of the change of base to $\{V_1, V_2, V_3\}$. As in case (1), $[\mathbf{P}]\,[\mathbf{A}]\,[\mathbf{P}'] = [\mathbf{B}]$ is symmetric. We also have $V_1 = (1,0,0)$, $V_2 = (0,1,0)$ and $V_3 = (0,0,1)$. Thus

$$\mathbf{B}V_1 = \lambda_1 V_1 = (\lambda_1,0,0)\ ,$$

$$\mathbf{B}V_2 = \lambda_2 V_2 = (0,\lambda_2,0)\ , \quad \text{and}$$

$$[\mathbf{B}] = \begin{pmatrix} \lambda_1 & 0 & 0 \\ 0 & \lambda_2 & 0 \\ k & l & m \end{pmatrix}. \quad \text{But } [\mathbf{B}] \text{ is}$$

symmetric, i.e., $k = l = 0$ and

$$[\mathbf{B}] = \begin{pmatrix} \lambda_1 & 0 & 0 \\ 0 & \lambda_2 & 0 \\ 0 & 0 & m \end{pmatrix}.$$

Further, $\mathbf{B}V_3 = (0\ 0\ 1)\,[\mathbf{B}] = (0\ 0\ m) = mV_3$, i.e., $m = \lambda_3$.

(3) Suppose λ is a triple root of the characteristic equation of **A**. Let V_1 be a characteristic vector of unit length corresponding to λ. Choose V_2 and V_3 so that $\{V_1, V_2, V_3\}$ is orthonormal. Let [**P**] be the matrix of the change base to $\{V_1, V_2, V_3\}$. As in (1), [**P**] [**A**] [**P'**] is symmetric. Since $V_1 = (1,0,0)$ and $\mathbf{B}V_1 = \lambda V_1 = (\lambda,0,0)$, and since [**B**] is symmetric, we have

$$[\mathbf{B}] = \begin{pmatrix} \lambda & 0 & 0 \\ 0 & l & m \\ 0 & m & k \end{pmatrix}.$$

Therefore the characteristic equation of the transformation will be

$$(x-\lambda) \begin{vmatrix} l-x & m \\ m & k-x \end{vmatrix} = 0, \text{ or}$$

$$(x-\lambda)\ [x^2 - (l+k)x + lk-m^2] = 0. \text{ But}$$

$x^2 - (l+k)x + lk - m^2 = 0$ has λ as a double root, i.e., $(l+k)^2 - 4(lk-m^2) = 0$.

This implies $(l-k)^2 + 4m^2 = 0$. Therefore $m = 0$ and $l = k = \lambda$. That is,

$$[\mathbf{B}] = \begin{pmatrix} \lambda & 0 & 0 \\ 0 & \lambda & 0 \\ 0 & 0 & \lambda \end{pmatrix}.$$

This proves the theorem.

5.11 **Quadratic forms and their reduction to canonical form:** In this section we suppose the bases are orthonormal. A quadratic form in three variables, x, y, z, is an expression

(1) $$Q = ax^2 + by^2 + cz^2 + 2fyz + 2gxz + 2hxy\ .$$

It is clear that

(2) $$Q = (x\ y\ z) \begin{pmatrix} a & h & g \\ h & b & f \\ g & f & c \end{pmatrix} \begin{pmatrix} x \\ y \\ z \end{pmatrix}.$$

For convenience let us denote the matrix

$$\begin{pmatrix} a & h & g \\ h & b & f \\ g & f & c \end{pmatrix}$$

by **A** and other matrices introduced by capital letters. Then we can write (2) as follows:

(3) $$Q = (x\ y\ z)\ \mathbf{I\ A\ I} \begin{pmatrix} x \\ y \\ z \end{pmatrix}.$$

Writing the unit matrix **I** as the product **SS'** where **S** is a unitary matrix, we get

(4) $$Q = (x\ y\ z)\ \mathbf{SS'}\ \ \mathbf{A}\ \ \mathbf{SS'} \begin{pmatrix} x \\ y \\ z \end{pmatrix}.$$

But by 2.9 we know that the products $(x\ y\ z)$ **S** and **S'** $\begin{pmatrix} x \\ y \\ z \end{pmatrix}$ represent the same vector. Suppose

$(x\ y\ z)\ \mathbf{S} = (X\ Y\ Z)$. Then

(5) $$Q = (X\ Y\ Z)\ \mathbf{S}'\ \mathbf{A}\ \mathbf{S} \begin{pmatrix} X \\ Y \\ Z \end{pmatrix}.$$

Since $\mathbf{A}$ is symmetric, by 5.10 we can find a unitary matrix $\mathbf{S}$ such that $\mathbf{S}'\ \mathbf{A}\ \mathbf{S} = \Lambda$, where

$$\Lambda = \begin{pmatrix} \lambda_1 & 0 & 0 \\ 0 & \lambda_2 & 0 \\ 0 & 0 & \lambda_3 \end{pmatrix}.$$

Consequently

(6) $$Q = (X\ Y\ Z) \begin{pmatrix} \lambda_1 & 0 & 0 \\ 0 & \lambda_2 & 0 \\ 0 & 0 & \lambda_3 \end{pmatrix} \begin{pmatrix} X \\ Y \\ Z \end{pmatrix} = \lambda_1 X^2 + \lambda_2 Y^2 + \lambda_3 Z^2 .$$

$\lambda_1 X^2 + \lambda_2 Y^2 + \lambda_3 Z^2$ is called the *canonical form* of Q. If λ_1, λ_2, and λ_3 are all positive, the quadratic form Q is called *positive*.

Illustration: Let $2x^2 + 4xy + 5y^2 = 1$ be the conic for which the coordinate system is to be rotated in order to get rid of the xy term. Clearly

$$(1) = (x\ y) \begin{pmatrix} 2 & 2 \\ 2 & 5 \end{pmatrix} \begin{pmatrix} x \\ y \end{pmatrix}$$

is the same conic section. Therefore

$$\begin{vmatrix} 2-\lambda & 2 \\ 2 & 5-\lambda \end{vmatrix} = 0$$

is the characteristic equation of its matrix and the proper values are 1 and 6. We have two choices:

(1) $$X^2 + 6Y^2 = 1 ,$$

(2) $$6X^2 + Y^2 = 1 .$$

Indeed there are two rotations of the base which transform the given equation to canonical form.

To find the matrices of these rotations we make use of the equations (5), (6), and (7) of 5.7. I.e.,

$$\begin{cases} (2-\lambda_1)d_{11} + 2d_{21} = 0 \\ 2d_{11} + (5-\lambda_1)d_{21} = 0 , \end{cases} \qquad \begin{cases} (2-\lambda_2)d_{12} + 2d_{22} = 0 \\ 2d_{12} + (5-\lambda_2)d_{22} = 0 \end{cases}$$

which give one of the rotations, and if we interchange λ_1 and λ_2 in the above systems of equations we get the other rotation. Here we have

$$(3) \quad \begin{cases} d_{11} + 2d_{21} = 0 \\ 2d_{11} + 4d_{21} = 0 , \end{cases} \qquad (4) \quad \begin{cases} -4d_{12} + 2d_{22} = 0 \\ 2d_{12} - d_{22} = 0 . \end{cases}$$

From (3) and (4) we get $d_{11} = -2d_{21}$ and $d_{22} = 2d_{12}$. Clearly we have many choices for d's but we choose them so that the matrix **D** will be unitary. Therefore

$$\mathbf{D} = \begin{pmatrix} \dfrac{-2\sqrt{5}}{5} & \dfrac{\sqrt{5}}{5} \\ \dfrac{\sqrt{5}}{5} & \dfrac{2\sqrt{5}}{5} \end{pmatrix}.$$

5.12 **Reduction to sum or differences of squares:** As in 5.11 the quadratic form $Q = ax^2 + by^2 + cz^2 + 2fyz + 2gxz + 2hxy$, by a change of base, can be reduced to $Q = \lambda_1 X^2 + \lambda_2 Y^2 + \lambda_3 Z^2$. The change of base is a unitary transformation.

If we consider the following transformation:

$$X\sqrt{|\lambda_1|} = \xi \text{ if } \lambda_1 \neq 0,$$

$$Y\sqrt{|\lambda_2|} = \eta \text{ if } \lambda_2 \neq 0,$$

$$Z\sqrt{|\lambda_3|} = \zeta \text{ if } \lambda_3 \neq 0,$$

we get $Q = \pm\,\xi^2 \pm \eta^2 \pm \zeta^2$, where the sign is the same as the sign of the corresponding λ.

This transformation is a change of base with matrix

$$\mathbf{T} = \begin{pmatrix} \dfrac{1}{\sqrt{|\lambda_1|}} & 0 & 0 \\ 0 & \dfrac{1}{\sqrt{|\lambda_2|}} & 0 \\ 0 & 0 & \dfrac{1}{\sqrt{|\lambda_3|}} \end{pmatrix}.$$

However, if, for example, $\lambda_3 = 0$, the term ζ^2 will be missing in the reduced form of Q. In this case the term $\dfrac{1}{\sqrt{|\lambda_3|}}$ in the matrix **T** above is meaningless. For this matrix to be a change of base we may write

$$\mathbf{T} = \begin{pmatrix} \dfrac{1}{\sqrt{|\lambda_1|}} & 0 & 0 \\ 0 & \dfrac{1}{\sqrt{|\lambda_2|}} & 0 \\ 0 & 0 & k \end{pmatrix}$$

for any non-zero scalar k. Note that in general the transformation **T** is not unitary.

5.13 **Simultaneous reduction of two quadratic forms:** Let

$$Q_1 = (x\ y\ z)\,\mathbf{A}\begin{pmatrix} x \\ y \\ z \end{pmatrix} \quad \text{and}$$

$$Q_2 = (x\ y\ z)\,\mathbf{B}\begin{pmatrix} x \\ y \\ z \end{pmatrix}$$

be two quadratic forms for which Q_1 is positive. Then we can reduce Q_1 to sum of squares and Q_2 to canonical form by a single change of base.

There exists a unitary transformation **U** which reduces Q_1 to the form $\lambda_1 X^2 + \lambda_2 Y^2 + \lambda_3 Z^2$, with λ_1, λ_2, and λ_3 positive. The change of base of 5.12, reducing $\lambda_1 X^2 + \lambda_2 Y^2 + \lambda_3 Z^2$ to a sum of squares is **T** which is symmetric. We have in fact

$$(X\ Y\ Z)\begin{pmatrix} \frac{1}{\sqrt{\lambda_1}} & 0 & 0 \\ 0 & \frac{1}{\sqrt{\lambda_2}} & 0 \\ 0 & 0 & \frac{1}{\sqrt{\lambda_3}} \end{pmatrix}\begin{pmatrix} \lambda_1 & 0 & 0 \\ 0 & \lambda_2 & 0 \\ 0 & 0 & \lambda_3 \end{pmatrix}\begin{pmatrix} \frac{1}{\sqrt{\lambda_1}} & 0 & 0 \\ 0 & \frac{1}{\sqrt{\lambda_2}} & 0 \\ 0 & 0 & \frac{1}{\sqrt{\lambda_3}} \end{pmatrix}\begin{pmatrix} X \\ Y \\ Z \end{pmatrix}$$

$$= (\xi\ \eta\ \zeta)\begin{pmatrix} 1 & 0 & 0 \\ 0 & 1 & 0 \\ 0 & 0 & 1 \end{pmatrix}\begin{pmatrix} \xi \\ \eta \\ \zeta \end{pmatrix},$$

and

$$\mathbf{T} = \begin{pmatrix} \frac{1}{\sqrt{\lambda_1}} & 0 & 0 \\ 0 & \frac{1}{\sqrt{\lambda_2}} & 0 \\ 0 & 0 & \frac{1}{\sqrt{\lambda_3}} \end{pmatrix}.$$

But

$$\begin{pmatrix} \lambda_1 & 0 & 0 \\ 0 & \lambda_2 & 0 \\ 0 & 0 & \lambda_3 \end{pmatrix}$$

is equal to $\mathbf{U\ A\ U'}$, since **U** is unitary. Therefore $\mathbf{T\ U\ A\ U'\ T} = \mathbf{I}$. Let $\mathbf{T\ U} = \mathbf{S}$. Then $\mathbf{S'} = \mathbf{U'T'} = \mathbf{U'T}$, and

(1) $\mathbf{S\ A\ S'} = \mathbf{I}$.

Applying the same change of base to **B**, we will have

$$Q_2 = (\xi\ \eta\ \zeta)\ \mathbf{B}_1 \begin{pmatrix} \xi \\ \eta \\ \zeta \end{pmatrix},$$

where $\mathbf{B}_1 = \mathbf{S\ B\ S'}$. Applying a unitary change of base does not change the new form of Q_1, since if $\mathbf{U}_1$ is unitary, $\mathbf{U}_1\mathbf{I}\ \mathbf{U}_1' = \mathbf{U}_1\mathbf{U}_1' = \mathbf{I}$. Let us choose $\mathbf{U}_1$ to change $\mathbf{B}_1$ to diagonal form. We get

$$Q_1 = \xi_1^{\ 2} + \eta_1^{\ 2} + \zeta_1^{\ 2},$$

$$Q_2 = \mu_1\ \xi_1^{\ 2} + \mu_2\ \eta_1^{\ 2} + \mu_3\ \zeta_1^{\ 2}.$$

But the characteristic equation of $\mathbf{B}_1$ is

(2) $|\mathbf{B}_1 - \mu\ \mathbf{I}| = 0 = |\mathbf{S\ B\ S'} - \mu\ \mathbf{I}|$.

Using (1) we get (2) in the form

$$|\mathbf{S}\,\mathbf{B}\,\mathbf{S}' - \mu\,\mathbf{S}\,\mathbf{A}\,\mathbf{S}'| = 0, \quad \text{or} \quad |\mathbf{S}(\mathbf{B} - \mu\,\mathbf{A})\mathbf{S}'| = 0\,.$$

This implies that

$$|\mathbf{S}|\;|\mathbf{B} - \mu\,\mathbf{A}|\;|\mathbf{S}'| = 0\,.$$

Since $|\mathbf{S}| = |\mathbf{S}'| = |\mathbf{T}|\,|\mathbf{U}| = \dfrac{1}{\sqrt{\lambda_1\,\lambda_2\,\lambda_3}} \neq 0$, we have

$$(3) \qquad |\mathbf{B} - \mu\,\mathbf{A}| = 0\,,$$

i.e., the μ's are the roots of (3).

Illustration: Change $5x^2 + 3y^2 + 3z^2 + 2xy - 2xz - 2yz$ to a sum or difference of squares, and find the change of base.

Here

$$Q = (x\;y\;z)\begin{pmatrix} 5 & 1 & -1 \\ 1 & 3 & -1 \\ -1 & -1 & 3 \end{pmatrix}\begin{pmatrix} x \\ y \\ z \end{pmatrix}.$$

The characteristic equation of the matrix of Q is

$$\begin{vmatrix} 5-\lambda & 1 & -1 \\ 1 & 3-\lambda & -1 \\ -1 & -1 & 3-\lambda \end{vmatrix} = 0, \quad \text{or}$$

$$(5-\lambda)\,(3-\lambda)^2 + 3\lambda - 9 = 0\,.$$

Hence $\qquad (3-\lambda)\,(\lambda^2 - 8\lambda + 12) = 0\,,$

and the characteristic values are 2, 3, and 6. Since these are all positive, Q reduces to $\xi^2 + \eta^2 + \zeta^2$.

Let the matrix of the change of base be $\mathbf{S} = \mathbf{TU}$, where $\mathbf{U}$ is unitary, and

$$\mathbf{T} = \begin{pmatrix} \dfrac{1}{\sqrt{2}} & 0 & 0 \\ 0 & \dfrac{1}{\sqrt{3}} & 0 \\ 0 & 0 & \dfrac{1}{\sqrt{6}} \end{pmatrix}.$$

By equation (5), (6), (7) of 5.7

$$\begin{cases} 3d_{11} + d_{21} - d_{31} = 0 \\ d_{11} + d_{21} - d_{31} = 0 \\ -d_{11} - d_{21} + d_{31} = 0\,, \end{cases} \quad \begin{cases} 2d_{12} + d_{22} - d_{32} = 0 \\ d_{12} - d_{32} = 0 \\ -d_{12} - d_{22} = 0\,, \end{cases} \quad \begin{cases} -d_{13} + d_{23} - d_{33} = 0 \\ d_{13} - 3d_{23} - d_{33} = 0 \\ -d_{13} - d_{23} - 3d_{33} = 0\,. \end{cases}$$

These equations give

$$d_{11} = 0, \; d_{21} = d_{31}, \; d_{12} = -d_{22} = d_{32}, \; d_{13} = 2d_{23} = -2d_{33}\,.$$

For the matrix **U** to be unitary, we choose

$$\begin{pmatrix} 0 & \frac{1}{\sqrt{2}} & \frac{1}{\sqrt{2}} \\ \frac{1}{\sqrt{3}} & \frac{-1}{\sqrt{3}} & \frac{1}{\sqrt{3}} \\ \frac{2}{\sqrt{6}} & \frac{1}{\sqrt{6}} & \frac{-1}{\sqrt{6}} \end{pmatrix}.$$

Then

$$\mathbf{S} = \mathbf{TU} = \begin{pmatrix} 0 & \frac{1}{2} & \frac{1}{2} \\ \frac{1}{3} & \frac{-1}{3} & \frac{1}{3} \\ \frac{1}{3} & \frac{1}{6} & \frac{-1}{6} \end{pmatrix}.$$

The reader may verify that $\mathbf{S\,A\,S'} = \mathbf{I}$.

EXERCISES 5

1. Find the characteristic values and characteristic vectors of the transformations of exercises 1, 3, 5, 7, 17 of chapter 2.
2. Show that the characteristic values of a unitary transformation have absolute value one.
3. Suppose **A** is a projection, on an axis or a plane through the origin.

   (i) Show that $\mathbf{A}^2 = \mathbf{A}$;
   (ii) Show that **A** is selfadjoint ;
   (iii) Find the characteristic values of **A** ;
   (iv) Change **A** to diagonal form.

4. Find the characteristic values of

   (i) $\begin{pmatrix} 2 & 0 \\ 1 & 1 \end{pmatrix}$, (ii) $\begin{pmatrix} 0 & 1 \\ 2 & 0 \end{pmatrix}$, (iii) $\begin{pmatrix} 0 & 2 & 3 \\ 3 & 2 & 5 \\ 1 & -1 & -2 \end{pmatrix}$, (iv) $\begin{pmatrix} 3 & 0 & 2 \\ 0 & 1 & -1 \\ 2 & -1 & 5 \end{pmatrix}$.

5. Let **A** and **B** be selfadjoint. Show that **AB** is selfadjoint if and only if **AB** = **BA**.
6. Find a change of base which reduces the following to diagonal form:

   (i) $\begin{pmatrix} 3 & 1 \\ 1 & 2 \end{pmatrix}$, (ii) $\begin{pmatrix} 2 & 3 \\ 3 & -5 \end{pmatrix}$, (iii) $\begin{pmatrix} 1 & -2 & 2 \\ -2 & 1 & 2 \\ 2 & 2 & -8 \end{pmatrix}$, (iv) $\begin{pmatrix} 1 & 0 & 1 \\ 0 & 0 & 2 \\ 1 & 2 & -1 \end{pmatrix}$

7. Reduce to canonical form

   (i) $8xy + 4y^2 = 1$, (ii) $4xy = 3$,
   (iii) $x^2 + 2xy + 4xz + 3y^2 + 7yz + 7z^2 = 15$, (iv) $3x^2 + 2y^2 + 3z^2 + 2xz = 5$.

8. Reduce the quadratic forms in exercise 7 to sum or difference of squares.

9. Show that $\mathbf{AA'}$ and $\mathbf{A'A}$ are selfadjoint for any $\mathbf{A}$.
10.* Show that the characteristic values of $\mathbf{A\,A'}$ are the same as those of $\mathbf{A'A}$, and they are all non-negative.
11. Find the matrices of the changes of base in exercises 7 and 8.
12. Reduce the following to canonical form:

(i) $52x^2 - 72xy + 73y^2$, (ii) $24xy - 7y^2$,

(iii) $17x^2 - 6xy + 9y^2$, (iv) $16x^2 + 24xy + 9y^2$,

(v) $x^2 + 4xy + 4y^2$.

13. Find the matrices of the changes of base in exercise 12.
14. Reduce to canonical form:

(i) $3x^2 + 2y^2 + 4z^2 - 4xy - 4xz$, (ii) $4y^2 - 2xy - xz + 2yz$,

(iii) $x^2 + y^2 - 8z^2 - 4xy + 4xz + 4yz$, (iv) $x^2 - z^2 + 2xz + 4yz$,

(v) $x^2 - 3y^2 + 3z^2 + 8yz$,

15. Find the matrices of the changes of base in exercise 14.
16. If the quadratic forms of exercises 12 and 14 are reduced to sum and difference of squares, find the matrices of changes of base.
17. In the following pairs of quadratic forms, the first is positive. Reduce them simultaneously, the first to a sum of squares and the second to canonical form:

(i) $17x^2 - 6xy + 9y^2$, $3x^2 + 8xy - 3y^2$, (ii) $66x^2 - 24xy + 59y^2$, $4xy$, (iii) $10x^2 + 13y^2 + 13z^2 - 4xy - 10yz - 4xz$, $2x^2 + 2y^2 + 3z^2 + 2xz + 2yz$.

18. Find the matrices of the changes of base in exercise 17.

ADDITIONAL PROBLEMS 5

1. Show that the proper values of

$$\begin{pmatrix} \frac{1}{2} & \frac{\sqrt{3}}{2} \\ \frac{-\sqrt{3}}{2} & \frac{1}{2} \end{pmatrix}, \quad \text{and} \quad \begin{pmatrix} \frac{\sqrt{2}}{2} & \frac{\sqrt{2}}{2} \\ \frac{-\sqrt{2}}{2} & \frac{\sqrt{2}}{2} \end{pmatrix}$$

are complex numbers with absolute values one.

2. Show that any rotation of the place about the origin through an angle θ has complex proper values with absolute value one except for $\theta = k\pi$, where k is an integer. Show that the rotation through $2k\pi$ is equivalent to $\mathbf{I}$.
3. Find the proper values and proper vectors of a symmetry in the plane with respect to:

(i) the origin, (ii) the x-axis, (iii) the line $ax + by = 0$.

4. Find the proper values and proper vectors of a projection of the plane on:

(i) the x-axis, (ii) the line $ax + by = 0$.

5. Find the proper values and proper vectors of the symmetry in the space with respect to

(i) the origin, (ii) the x-axis, (iii) the xy-plane, (iv) the line: $\begin{cases} x = t \\ y = 2t \\ z = 3t \end{cases}$,

(v) the plane $ax + by + cz = 0$.

6. Find the proper values and proper vectors of the projection on:

(i) $x + y + z = 0$, (ii) $ax + by + cz = 0$.

7. In problems 3, 4, 5, 6, if possible, change the matrices of the transformations to diagonal form.
8. Find the matrices of the changes of base in exercise 7.
9. Find the proper values and proper vectors of

(i) $\begin{pmatrix} 1 & 0 & 0 \\ 0 & \sin\theta & \cos\theta \\ 0 & -\cos\theta & \sin\theta \end{pmatrix}$, (ii) $\begin{pmatrix} \sin\theta\cos\phi & \sin\theta\sin\phi & \cos\theta \\ \cos\theta\cos\phi & \cos\theta\sin\phi & -\sin\theta \\ \sin\phi & -\cos\phi & 0 \end{pmatrix}$.

10. Find the proper values and proper vectors of

$$\begin{pmatrix} \frac{2}{3} & \frac{1}{3} & \frac{-2}{3} \\ \frac{1}{3} & \frac{2}{3} & \frac{2}{3} \\ \frac{2}{3} & \frac{-2}{3} & \frac{1}{3} \end{pmatrix}, \quad \text{and} \quad \begin{pmatrix} \frac{1}{\sqrt{6}} & \frac{-1}{\sqrt{6}} & \frac{2}{\sqrt{6}} \\ \frac{1}{\sqrt{2}} & \frac{1}{\sqrt{2}} & 0 \\ \frac{-1}{\sqrt{3}} & \frac{1}{\sqrt{3}} & \frac{1}{\sqrt{3}} \end{pmatrix}.$$

11. Show that in the space any unitary transformation has a real proper value which is ± 1.
12. Show that in the space under a unitary transformation there is a straight line which is transformed to itself. (In case the transformation is a rotation this line is called the axis of rotation).
13. Reduce to canonical form:

(i) $3x^2 - xy + y^2$, (ii) $x^2 + 2xy + 3y^2$,

(iii) $x^2 + 4xy + y^2$, (iv) $25x^2 + 36xy + 40y^2$.

14. Find the matrices of the changes of base in problem 13.
15. Reduce to canonical form:

(i) $16x^2 + 4y^2 + 9z^2 - 16xy + 24xz - 12yz$, (ii) $2x^2 - y^2 - z^2 + 2xy + 2xz + 4yz$,

(iii) $5x^2 + 3y^2 + 3z^2 + 2xy - 2xz - 2yz$, (iv) $3x^2 + 2xy + 2xz + 4yz$,

(v) $2y^2 - z^2 - 4yz$.

16. Find the matrices of the changes of base in problem 15. Find the equation of the invariant line in each case.
17. Show that the characteristic vectors of a projection on a plane $ax + by + cz = 0$ corresponding to non-zero characteristic values are in this plane.
18. Show that the characteristic vectors of a projection on an axis

$$\begin{cases} x = at \\ y = bt \\ z = ct \end{cases}$$

which correspond to non-zero characteristic values are on this line.
19. Show that the characteristic vectors of a symmetry with respect to the plane $ax + by + cz = 0$ are in this plane or are orthogonal to it.
20. Show that the characteristic vectors of a symmetry with respect to an axis are on that axis or orthogonal to it.
21. What can be said about the characteristic vectors of the symmetry with respect to the origin?

PART II

6. UNITARY SPACES

Introduction As we have observed in the first chapter, we may introduce into a Euclidean space a coordinate system, in such a way that, for example, to every point or vector in the plane we may make correspond a pair (x,y) of real numbers. Similarly to every vector in the space we may associate an ordered triple (x,y,z) of real numbers. Two immediate simple generalizations of these ideas may be introduced. First we may consider the set of all ordered sets of n real numbers for some fixed integer n, with a definition of addition of two such sets, and of multiplication of a set by a number. The set is then called a *real space.* The second generalization is to the notion of a *complex space.* That is, we may consider the set of all ordered sets of n complex numbers with appropriate definitions for addition and multiplication by a scalar as given below. This set is called *complex space.*

It is not the purpose of this book to develop all results in detail. We hope rather to outline the major results in reasonable sequence, sketch a few important proofs, and indicate intuitively the method of proving some other theorems. The applications of the theory of linear transformations discussed earlier will also be generalized, and the techniques employed will be applied to some special geometric problems.

6.1 **Scalars, Vectors and vector spaces:** Let any complex number be called a *scalar.* In what follows we will use Greek letters for vectors, and a *vector* ξ is defined as an ordered n-tuple of scalars $(x_1, \ldots, x_n)$ having the following properties:

1. For any two vectors $\xi = (x_1, \ldots, x_n)$ and $\eta = (y_1, \ldots, y_n)$ we define the sum to be $\xi + \eta = (x_1 + y_1, \ldots, x_n + y_n)$.
2. For any scalar a and any vector $\xi = (x_1, \ldots, x_n)$ we define

$$a\,\xi = (ax_1, \ldots, ax_n)\,.$$

The set $(0, \ldots, 0) = 0$ is called the zero vector.

The set of all these vectors is called a *vector space* V_n over the complex numbers. The symbol V_n will always be used to refer to this set in what follows.

6.2 **Subspaces:** Any subset S of V_n, for which the sum of any two vectors of S is a vector in S and the product of a vector of S by a scalar is also in S, is called a *subspace* of V_n. These subspaces are also called vector spaces.

6.3 **Linear independence:** A set of vectors $\{\xi_1, \ldots, \xi_k\}$ is called *linearly independent* if for scalars $c_1, \ldots, c_k$,

$$c_1\xi_1 + \ldots + c_k\,\xi_k = 0 \quad \text{if and only if}$$

$$c_1 = \ldots = c_k = 0\,; \quad \text{otherwise} \quad \{\xi_1, \ldots, \xi_k\}$$

is called *linearly dependent.* It should be noted that any set of vectors containing the zero vector is linearly dependent.

Clearly any subset of a linearly independent set of vectors is also linearly independent. We leave the proof to the reader as an exercise.

6.4 **Theorem:** The vectors $\xi_1, \ldots, \xi_n$ are linearly dependent if and only if some ξ_i is a linear combination of the others.

Proof: Let $c_1\,\xi_1 + \ldots + c_n\,\xi_n = 0$. If $\{\xi_1, \ldots, \xi_n\}$ is linearly dependent, then a set $c_1, c_2, \ldots, c_n$

can be found for which some c_i, say c_p, is different from zero. Therefore

$$\xi_p = -\frac{c_1}{c_p}\xi_1 - \ldots - \frac{c_{p-1}}{c_p}\xi_{p-1} - \frac{c_{p+1}}{c_p}\xi_{p+1} - \ldots - \frac{c_n}{c_p}\xi_n \; .$$

Conversely, let

$$\xi_p = d_1\,\xi_1 + \ldots + d_{p-1}\,\xi_{p-1} + d_{p+1}\,\xi_{p+1} + \ldots + d_n\,\xi_n \; .$$

Then $$d_1\,\xi_1 + \ldots + d_{p-1}\,\xi_{p-1} + (-1)\,\xi_p + d_{p+1}\,\xi_{p+1} + \ldots + d_n\,\xi_n = 0 \; ,$$

and this proves the theorem.

6.5 **Base:** A set of vectors $\{\epsilon_1, \ldots, \epsilon_n\}$ is called a *base* for a vector space V_n if:

(1) $\epsilon_1, \ldots, \epsilon_n$ are linearly independent.

(2) Any vector of V_n can be written as a linear combination of $\epsilon_1, \ldots, \epsilon_n$. We say $\{\epsilon_1, \ldots, \epsilon_n\}$ generates the space V_n. It is an exercise for the reader to show that each vector of V_n can be written in terms of $\epsilon_1, \ldots, \epsilon_n$ in one and only one way. The ordered set of the coefficients of $\epsilon_1, \ldots, \epsilon_n$ are called the *components* of the vector with respect to that base.

6.6 **Theorem:** Let $\{\alpha_1, \ldots, \alpha_k\}$ be any set of vectors. Let $\beta_1, \ldots, \beta_{k+1}$ be vectors, each a linear combination of $\alpha_1, \ldots, \alpha_k$. Then $\beta_1, \ldots, \beta_{k+1}$ are linearly dependent.

Proof: If $\beta_1 = 0$, since $1.\ \beta_1 + 0\ \beta_2 + \ldots + 0\ \beta_{k+1} = 0$, the set is linearly dependent. Otherwise we may write $\beta_1 = a_{11}\ \alpha_1 + a_{12}\ \alpha_2 + \ldots + a_{1k}\ \alpha_k$, and there is some $a_{1i} \neq 0$. Then

$$\alpha_i = -\frac{1}{a_{1i}}\beta_1 - \frac{a_{11}}{a_{1i}}\alpha_1 - \ldots - \frac{a_{1,i-1}}{a_{1i}}\alpha_{i-1} - \frac{a_{1,i+1}}{a_{1i}}\alpha_{i+1} - \ldots - \frac{a_{1k}}{a_{1i}}\alpha_k \; .$$

Then any linear combination of $\alpha_1, \ldots, \alpha_k$ is also a linear combination of the vectors $\beta_1, \alpha_1, \ldots, \alpha_{i-1}, \alpha_{i+1}, \ldots, \alpha_k$. In particular

$$\beta_2 = b_{21}\,\beta_1 + a_{21}\ \alpha_1 + \ldots + a_{2,i-1}\ \alpha_{i-1} + a_{2,i+1}\ \alpha_{i+1} + \ldots + a_{2k}\ \alpha_k \; .$$

If every a_{2j}, $j = 1, \ldots, i-1, i+1, \ldots, k$, is zero, $\beta_2 = b_{21}\,\beta_1$, and the set $\{\beta_1, \ldots, \beta_{k+1}\}$ is linearly dependent. If some $a_{2j} \neq 0$, then as above α_j is a linear combination of β_1, β_2 and the $k-2$ vectors remaining of the original set. Again any vector which is a linear combination of $\beta_1, \alpha_1, \ldots, \alpha_{i-1}, \alpha_{i+1}, \ldots, \alpha_k$ hence also of $\alpha_1, \ldots, \alpha_k$ is a linear combination of the vectors β_1, β_2 and the $k-2$ vectors remaining of the original set of α's, so that in particular

$$\beta_3 = b_{31}\,\beta_1 + b_{32}\,\beta_2 + a_{31}\ \alpha_1 + \ldots + a_{3k}\ \alpha_k \; ,$$

where α_i and α_j are omitted. Once more either every a_{3l}, $l = 1, \ldots, k$, $l \neq i$, $l \neq j$, is zero, so that $\beta_3 = b_{31}\,\beta_1 + b_{32}\,\beta_2$ or some α_l is a linear combination of $\beta_1, \beta_2, \beta_3$, and the $k-3$ remaining vectors of the set $\{\alpha_1, \ldots, \alpha_k\}$.

Continuing in this way, eliminating the α's and adding β's one at a time, at each stage either every a_{mn} is zero, and β_m is a linear combination of $\beta_1, \ldots, \beta_{m-1}$ or some $a_{mn} \neq 0$, and α_n is a linear combination of $\beta_1, \ldots, \beta_m$, and the $k-m$ remaining vectors of the original set $\{\alpha_1, \ldots, \alpha_k\}$. Thus either for some $m \leq k$,

$$\beta_m = b_1\,\beta_1 + \ldots + b_{m-1}\,\beta_{m-1} \; ,$$

or else $$\beta_{k+1} = c_1\,\beta_1 + c_2\,\beta_2 + \ldots + c_{k-1}\,\beta_{k-1} + c_k\,\beta_k \; .$$

Thus in any case $\beta_1, \ldots, \beta_{k+1}$ are linearly dependent.

6.7 **Dimension theorem:** Any base for the space V_n has exactly n elements.

Proof: We observe that the n vectors $\epsilon_i = (x_1, \ldots, x_n)$, $i = 1, \ldots, n$ having $x_i = 1$, $x_j = 0$ for $j \neq i$, form a base for V_n. That is, for any vector $(y_1, \ldots, y_n)$ we have

$$(y_1, \ldots, y_n) = y_1 \epsilon_1 + \ldots + y_n \epsilon_n, \quad \text{while if}$$

$$z_1 \epsilon_1 + \ldots + z_n \epsilon_n = 0, \quad \text{then} \quad z_1 = \ldots = z_n = 0. \quad \text{We consider}$$

now any base $\{\alpha_1, \ldots, \alpha_k\}$ for V_n.

We observe that k cannot exceed n, for if $k \geq n + 1$, then, by 6.6 the vectors $\alpha_1, \ldots, \alpha_{n+1}$ are linearly dependent. Further, since every vector of V_n would have to be a linear combination of the vectors of $\{\alpha_1, \ldots, \alpha_k\}$, in particular each ϵ_i would be such a linear combination. If k were less than n, then by 6.6, $\epsilon_1, \ldots, \epsilon_n$ would be linearly dependent. Thus $k = n$.

Since the number of elements of any base for V_n *is n we say* V_n *is of dimension n.*

The reader may show that for any vector space as defined in 6.1 or 6.2, if there is a base with n elements any other base contains exactly n elements. Thus *the number of elements in a base for a vector space is called the dimension of that space.*

6.8 **Inner Product:** Let $\xi = (x_1, \ldots, x_n)$ and $\eta = (y_1, \ldots, y_n)$ be two vectors. The *inner product* of ξ and η in that order is defined as follows:

$$(\xi, \eta) = x_1\bar{y}_1 + \ldots + x_n\bar{y}_n,$$

where by the symbol $\bar{z}$ we mean the complex conjugate of the complex number z.

Here the inner product is not commutative, i.e.

$$(\xi, \eta) \neq (\eta, \xi). \quad \text{In fact} \quad (\eta, \xi) = \overline{(\xi, \eta)}$$

6.9 **Unitary spaces:** A *unitary* space or a *Euclidean* space is a vector space V_n with inner product defined. We denote an n-dimensional unitary space by E_n.

6.10 **Definition:** The *norm* of a vector $\xi = (x_1, \ldots, x_n)$ is defined to be $|\xi| = (\xi, \xi)^{\frac{1}{2}}$.

Clearly $$|\xi| = (x_1 \bar{x}_1 + \ldots + x_n \bar{x}_n)^{\frac{1}{2}} = (|x_1|^2 + \ldots + |x_n|^2)^{\frac{1}{2}}.$$

6.11 **Theorem:** Inner multiplication is distributive with respect to vector addition. That is, $(\alpha, [\beta_1 + \ldots + \beta_k]) = (\alpha, \beta_1) + \ldots + (\alpha, \beta_k)$.

Proof: Let $\alpha = (x_1, \ldots, x_n)$ and $\beta_i = (y_{i1}, \ldots, y_{in})$, $i = 1, \ldots, k$. Therefore

$$(\alpha, [\beta_1 + \ldots + \beta_k]) = x_1 (\bar{y}_{11} + \ldots + \bar{y}_{k1}) + \ldots + x_n (\bar{y}_{1n} + \ldots + \bar{y}_{kn}) =$$

$$(x_1 \bar{y}_{11} + \ldots + x_n \bar{y}_{1n}) + \ldots + (x_1 \bar{y}_{k1} + \ldots + x_n \bar{y}_{kn}) = (\alpha, \beta_1) + \ldots + (\alpha, \beta_k).$$

6.12 **Definition:** In E_n, two vectors ξ and η are said to be *orthogonal* if $(\xi, \eta) = 0$. Note that $(\xi, \eta) = 0$ implies $(\eta, \xi) = 0$.

6.13 **Theorem:** In E_n any set of orthogonal non-zero vectors is linearly independent.

Proof: Let $\{\alpha_1, \ldots, \alpha_k\}$ be a set of orthogonal non-zero vectors, i.e., $(\alpha_i, \alpha_j) = 0$ for $i \neq j$.

Let

$$(1) \qquad c_1 \alpha_1 + \ldots + c_k \alpha_k = 0.$$

Take the inner product of both sides of (1) by α_i. We know

$$(\alpha_i, \alpha_p) = 0 \quad \text{if} \quad i \neq p, \quad \text{while} \quad (\alpha_p, \alpha_p) = |\alpha_p|^2 \neq 0$$

Therefore we get

$$c_p\,|\alpha_p|^2 = 0 \quad \text{which implies} \quad c_p = 0 \quad \text{for all} \quad p = 1, \ldots, k\,.$$

This proves the theorem.

6.14 **Definition:** A set of vectors $\{\alpha_1, \ldots, \alpha_k\}$ in E_n is called *orthonormal* if $(\alpha_i, \alpha_j) = 0$ for $i \neq j$ and $(\alpha_i, \alpha_i) = 1$, where $i, j = 1, \ldots, k$.

6.15 **Orthonormalization of a set of vectors:** (Gram, Schmidt): Let $\{\xi_1, \ldots, \xi_k\}$ be a linearly independent set of vectors in E_n. Then we can construct an orthonormal set of vectors in E_n from $\{\xi_1, \ldots \xi_k\}$, which generates the same subspace of E_n.

Let $$\alpha_1 = \frac{\xi_1}{|\xi_1|} \qquad \text{Clearly } |\alpha_1| = 1\,.$$

Let $$\alpha_2 = \frac{\xi_2 - (\xi_2, \alpha_1)\,\alpha_1}{|\xi_2 - (\xi_2, \alpha_1)\,\alpha_1|}\,,$$

$\ldots \quad \ldots \quad \ldots$

$$\alpha_{p+1} = \frac{\xi_{p+1} - [(\xi_{p+1}, \alpha_1)\,\alpha_1 + \ldots + (\xi_{p+1}, \alpha_p)\,\alpha_p]}{|\xi_{p+1} - [(\xi_{p+1}, \alpha_1)\,\alpha_1 + \ldots + (\xi_{p+1}, \alpha_p)\,\alpha_p]|}\,, \quad p = 1, \ldots, k-1\,.$$

It is left to the reader to show that $\{\alpha_1, \ldots, \alpha_k\}$ is an orthonormal set, and it generates the same subspace generated by $\{\xi_1, \ldots, \xi_k\}$.

6.16 **Orthonormal base:** A set $\{\alpha_1, \ldots, \alpha_n\}$ is called an *orthonormal base* if $\{\alpha_1, \ldots, \alpha_n\}$ is a base and $\{\alpha_1, \ldots, \alpha_n\}$ is orthonormal. The reader may show that the components of any vector ξ with respect to the orthonormal base $\{\alpha_1, \ldots, \alpha_n\}$ are $(\xi, \alpha_1), \ldots, (\xi, \alpha_n)$.

6.17 **Theorem:** In any real unitary space

I $\quad |c\,\xi| = |c|\,|\xi|\,,$

II $\quad |\xi| \geqq 0$ and $|\xi| = 0$ if and only if $\xi = 0$,

III $\quad |(\xi, \eta)| \leqq |\xi|\,|\eta|\,;$ this is called the Schwarz inequality,

IV $\quad |\xi + \eta| \leqq |\xi| + |\eta|\,;$ this is called the triangle inequality.

Proof of III: In the spaces of three or less dimensions III is a direct result of the definition of inner product. Now in general, regardless of the dimension of the space, we observe that for two vectors α and β of unit norm

$$|\alpha - \beta|^2 = (\alpha - \beta, \alpha - \beta) = 1 + 1 - 2\,(\alpha, \beta) \geqq 0\,.$$

Therefore

$$1 \geqq (\alpha, \beta)\,.$$

Substituting $-\beta$ for β, we get

$$|\alpha|^2 + |\beta|^2 = 2 \geqq -2\,(\alpha, \beta)\,.$$

Thus

$$|(\alpha, \beta)| \leqq 1\,.$$

Now let $$\alpha = \frac{\xi}{|\xi|} \text{ and } \beta = \frac{\eta}{|\eta|}, \quad \xi, \eta \neq 0\,. \quad \text{Then}$$

$$\frac{|(\xi, \eta)|}{|\xi|\,|\eta|} \leqq 1\,, \quad \text{or}$$

$$|(\xi, \eta)| \leqq |\xi|\,|\eta|\,.$$

For the complete proof of all parts of the theorem in the complex space, see 10.17.

EXERCISES 6

1. Show that in a three-dimensional space the subspace $z = 0$ may be generated by the pair of vectors

   (i) $\{(1,0,0), (1,1,0)\}$ (ii) $\{(2,2,0), (4,1,0)\}$ (iii) $\{(5,-2,0), (-1,-1,0)\}$.

2. Show that the set of all polynomials of degree $n-1$ with real coefficients is a real vector space of dimension n and $\{1, x, \ldots, x^{n-1}\}$ is a base for this space, if to the polynomial $a_0 + a_1 x + \ldots + a_{n-1} x^{n-1}$ we associate the vector $(a_0, a_1, \ldots, a_{n-1})$.
3. Examine for linear dependence

   (i) $\{(1,i,1+i), (i,-1,2-i), (0,0,3)\}$ (ii) $\{(5,6,7,7), (2,0,0,0), (0,1,0,0), (0,-1,-1,0)\}$

4. Find two vectors which generate the subspace of all vectors

   (x_1,x_2,x_3,x_4) satisfying $x_1 + x_2 = x_3 - x_4 = 0$.

5. Show that each of the following sets of vectors is linearly independent.

   (i) $\{(3,-1,2,-3), (1,1,2,0), (3,-1,6,-6)\}$, (ii) $\{(0,-2,0,-2), (1,0,1,0)\}$.

6. If the two subspaces generated by the two sets of vectors in 5 are respectively called S and T, find the dimension of the subspace of the common part of S and T.
7. Let $\xi = (1,3,5,9)$ and $\eta = (0,-i,2,-1)$

   (i) find $|\xi|$ and $|\eta|$,
   (ii) find (ξ, η) ,
   (iii) orthonormalize $\{\xi, \eta\}$.

8.* Write the normal equation of a hyperplane in n-dimensional space.
9.* Show that the common part of two subspaces of V_n is also a subspace. But the set consisting of all vectors of two subspaces of V_n is not necessarily a subspace.
10. Let k be a fixed positive integer less than n. Show that the set of vectors of the form $(x_1, \ldots, x_k, 0, \ldots, 0)$ is a subspace of V_n.
11. Show that the subspace in 10 is of dimension k.
12. Find the components of $(7,-2,1,1)$ with respect to the base

    $\{(1,1,0,0), (0,0,1,1), (1,0,0,1), (0,1,1,0)\}$.

13. In a right triangle let a and b be the sides of the right angle and h the altitude to the hypotenuse c. Show that

    (i) $$\frac{1}{h^2} = \frac{1}{a^2} + \frac{1}{b^2}\,,$$

    (ii) $h^2 =$ (projection of a on c) . (projection of b on c) ,

    using inner product of vectors.
14. In the right triangle of exercise 13, let d be the length of the angle bisector of the right angle. Show that

    $$\frac{\sqrt{2}}{d} = \frac{1}{a} + \frac{1}{b}$$

    using the inner product.

15. In a right triangle **ABC** let $\mathbf{C} = \frac{\pi}{2}$, $\mathbf{AC} = b$, $\mathbf{BC} = a$, and the altitude through **C** be $\mathbf{CH} = h$. Let the feet of the perpendiculars through **H** to **AC** and **BC** be respectively **E** and **F**. Show that

$$\frac{\mathbf{AE}}{\mathbf{BF}} = \frac{b^3}{a^3} .$$

16. Let $\{\eta_1, \eta_2, \eta_3\}$ be an orthogonal set of vectors in a real three-dimensional unitary space. Let $\xi = a\,\eta_1 + b\,\eta_2 + c\,\eta_3$ such that $a+b+c = 1$. Show that

(i) ξ is perpendicular to the plane passing through the end points of η_1, η_2, and η_3 if and only if

$$\frac{1}{|\xi|^2} = \frac{1}{|\eta_1|^2} + \frac{1}{|\eta_2|^2} + \frac{1}{|\eta_3|^2} .$$

(ii) If ξ makes equal angles with η_1, η_2, and η_3, then

$$\frac{\sqrt{3}}{|\xi|} = \frac{1}{|\eta_1|} + \frac{1}{|\eta_2|} + \frac{1}{|\eta_3|} .$$

17. Generalize exercise 16 to k vectors in E_n, $k \leqq n$.

18. Given two planes

(1) $\qquad a_1x + b_1y + c_1z + d_1 = 0$, and

(2) $\qquad a_2x + b_2y + c_2z + d_2 = 0$

show that the line of intersection of (1) and (2) has the direction numbers

$$l = \begin{vmatrix} b_1 & c_1 \\ b_2 & c_2 \end{vmatrix} , \quad m = \begin{vmatrix} c_1 & a_1 \\ c_2 & a_2 \end{vmatrix} , \quad n = \begin{vmatrix} a_1 & b_1 \\ a_2 & b_2 \end{vmatrix} .$$

19. Supply proofs of I, II, and IV of 6.17.

20. Let $\{\alpha_1, \alpha_2, \ldots, \alpha_k\}$ be an orthonormal set of vectors in a real unitary space. Show that for any vector ξ in the space,

$$(\xi, \xi) \geqq (\xi, \alpha_1)^2 + (\xi, \alpha_2)^2 + \ldots + (\xi, \alpha_k)^2 .$$

7. LINEAR TRANSFORMATIONS, MATRICES AND DETERMINANTS

7.1 **Definition:** A *linear transformation* **A** on V_n is a method of corresponding to each vector ξ of V_n another vector $\xi_1 = \mathbf{A}\,\xi$ of V_n such that for any two vectors ξ and η of V_n and scalars a and b,

$$\mathbf{A}\,(a\,\xi + b\,\eta) = a\mathbf{A}\,\xi + b\mathbf{A}\,\eta\ .$$

All transformations referred to will be linear.

For addition and multiplication of transformations we follow the definition in 2.2. The theorem 2.3 is also true for any vector space. The same proof carries over, since it is independent of the dimension of the space.

7.2 **Matrix of a Transformation A:** Let $\{\alpha_1, \ldots, \alpha_n\}$ be a base for V_n. Suppose

$$\mathbf{A}\,\alpha_1 = a_{11}\ \alpha_1 + \ldots + a_{1n}\ \alpha_n\,,$$

$$\mathbf{A}\,\alpha_2 = a_{21}\ \alpha_1 + \ldots + a_{2n}\ \alpha_n\,,$$

$$\ldots \qquad \ldots$$

$$\mathbf{A}\,\alpha_n = a_{n1}\ \alpha_1 + \ldots + a_{nn}\ \alpha_n\ .$$

We see that $(a_{11}, \ldots, a_{1n}), \ldots, (a_{n1}, \ldots, a_{nn})$ are respectively the components of the vectors $\mathbf{A}\alpha_1, \ldots, \mathbf{A}\alpha_n$ with respect to the base $\{\alpha_1, \ldots, \alpha_n\}$. Putting these components in n rows as follows:

$$\begin{pmatrix} a_{11} & \ldots & a_{1n} \\ a_{21} & \ldots & a_{2n} \\ \ldots & & \ldots \\ a_{n1} & \ldots & a_{nn} \end{pmatrix},$$

we call this the matrix of **A** with respect to $\{\alpha_1, \ldots, \alpha_n\}$. Here a_{ij} means the element in the i-th row and j-th column. The zero transformation **O** has a matrix with elements all zero. The identity transformation **I** has $a_{ii} = 1$, $i = 1, \ldots n$, and $a_{ij} = 0$, $i \neq j$.

7.3 **Addition and Multiplication of Matrices:** As in 2.6 addition of matrices can be defined in correspondence with the sum of transformations. The method is the same as in 2.7 and it is left to the reader to carry out the process.

For multiplication of matrices we also use the same definition as in 2.7. We shall demonstrate how the definition is justified. Let

$$\begin{pmatrix} a_{11} & \ldots & a_{1n} \\ \ldots & & \ldots \\ a_{n1} & \ldots & a_{nn} \end{pmatrix} \quad \text{and} \quad \begin{pmatrix} b_{11} & \ldots & b_{1n} \\ \ldots & & \ldots \\ b_{n1} & \ldots & b_{nn} \end{pmatrix} \quad \text{be}$$

the matrices of the transformations **A** and **B** with respect to the base $\{\alpha_1, \ldots, \alpha_n\}$. Let $\mathbf{C} = \mathbf{AB}$. It is observed that

$$\mathbf{A}\,\alpha_1 = a_{11}\ \alpha_1 + \ldots + a_{1n}\ \alpha_n\,,$$

$$\ldots \qquad \ldots \qquad \ldots$$

$$\mathbf{A}\alpha_n = a_{n1}\ \alpha_1 + \ldots + a_{nn}\ \alpha_n\ .$$

We easily see

$$\mathbf{B}(\mathbf{A}\ \alpha_1) = (a_{11}b_{11} + \ldots + a_{1n}b_{n1})\ \alpha_1 + \ldots + (a_{11}b_{1n} + \ldots + a_{1n}b_{nn})\ \alpha_n$$

$$\ldots \qquad \ldots \qquad \ldots$$

$$\mathbf{B}(\mathbf{A}\ \alpha_n) = (a_{n1}b_{11} + \ldots + a_{nn}b_{n1})\ \alpha_1 + \ldots + (a_{n1}b_{1n} + \ldots + a_{nn}b_{nn})\ \alpha_n\ .$$

Therefore we define the matrix of **AB** to be

$$\begin{pmatrix} a_{11}b_{11} + \ldots + a_{1n}b_{n1} & \ldots & a_{11}b_{1n} + \ldots + a_{1n}b_{nn} \\ \ldots & \ldots & \ldots \\ a_{n1}b_{11} + \ldots + a_{nn}b_{n1} & \ldots & a_{n1}b_{1n} + \ldots + a_{nn}b_{nn} \end{pmatrix} .$$

As was mentioned in 2.7 the multiplication of matrices or transformations is not commutative, although **O** and **I** commute with all transformations. But theorem 2.3 assures that the multiplication is associative, and multiplication is distributive with respect to addition.

7.4 **Rectangular matrices:** The definition in 2.8 is valid here. The matrix multiplication is the same as described in 2.8.

The most important rectangular matrices are the row or column matrices, which are vectors presented in matrix form.

The inner product of $\xi = (x_1, \ldots, x_n)$ and $\eta = (y_1, \ldots, y_n)$ in terms of matrix multiplications is

$$(\xi, \eta) = (x_1 \ldots x_n) \begin{pmatrix} \bar{y}_1 \\ \cdot \\ \cdot \\ \cdot \\ \bar{y}_n \end{pmatrix} = (x_1\bar{y}_1 + \ldots + x_n\bar{y}_n)\ .$$

The transform of a vector can also be represented as either

$$(1) \qquad \mathbf{A}\ \xi = (X_1 \ldots X_n) = (x_1 \ldots x_n) \begin{pmatrix} a_{11} & \ldots & a_{1n} \\ \ldots & & \ldots \\ a_{n1} & \ldots & a_{nn} \end{pmatrix},$$

or

$$(2) \qquad \mathbf{A}\ \xi = \begin{pmatrix} X_1 \\ \cdot \\ \cdot \\ \cdot \\ X_n \end{pmatrix} = \begin{pmatrix} a_{11} & \ldots & a_{n1} \\ \ldots & & \ldots \\ a_{1n} & \ldots & a_{nn} \end{pmatrix} \begin{pmatrix} x_1 \\ \cdot \\ \cdot \\ \cdot \\ x_n \end{pmatrix} .$$

Note that the matrix in (2) is the transpose of the matrix in (1).

7.5 **Determinants:** In 3.1 the determinant of a three-by-three matrix was defined. Here we suppose that the determinant of an $(n-1)$ – by – $(n-1)$ matrix is defined and we define the determinant of an n–by–n matrix as follows:

Let $$[\mathbf{A}] = \begin{pmatrix} a_{11} & \ldots & a_{1n} \\ \ldots & & \ldots \\ a_{n1} & \ldots & a_{nn} \end{pmatrix} .$$

We denote the determinant of [**A**], written det [**A**], by

$$\begin{vmatrix} a_{11} & \cdots & a_{1n} \\ \cdots & & \cdots \\ a_{1n} & \cdots & a_{nn} \end{vmatrix} .$$

The cofactor $\mathbf{A}_{ij}$ of the element a_{ij} of [**A**] is defined to be

$$\mathbf{A}_{ij} = (-1)^{(i+j)} \det [\mathbf{B}] ,$$

where [**B**] is the $(n-1)$ – by – $(n-1)$ matrix obtained from [**A**] by removing the row and column containing a_{ij}. Now we define

$$\det [\mathbf{A}] = a_{1j}\mathbf{A}_{1j} + \ldots + a_{nj}\mathbf{A}_{nj}, \quad \text{or}$$

$$\det [\mathbf{A}] = a_{i1}\mathbf{A}_{i1} + \ldots + a_{in}\mathbf{A}_{in} .$$

It is necessary to show that the result is independent of the choice of i or j. We leave it to the reader to show that in any case det [**A**] is the sum of all products $(-1)^p \; a_{1i_1} \cdot \ldots \cdot a_{ni_n}$, where $\{i_1, \ldots, i_n\}$ is a permutation of the integers 1, 2, . . . , n, and p is the number of times a larger i_j precedes a smaller one in the sequence $i_1, \ldots, i_n$. For example, in the sequence 1, 3, 5, 2, 4, we have 3 preceding 2, 5 preceding 2 and 4, hence $p = 3$. The properties mentioned in 3.2 are true for any n–by–n determinant. We also mention the following:

(1) $\det [\mathbf{O}] = 0$,

(2) $\det [\mathbf{I}] = 1$,

(3) $\det [\mathbf{AB}] = \det [\mathbf{A}] \cdot \det [\mathbf{B}] = \det [\mathbf{BA}]$.

We easily see that the following suggestion leads to a proof for (3).

The $2n$–by–$2n$ determinant

$$\begin{vmatrix} [\mathbf{A}] & [\mathbf{O}] \\ -[\mathbf{I}] & [\mathbf{B}] \end{vmatrix} = \det [\mathbf{A}] \det [\mathbf{B}] .$$

Also

$$\begin{vmatrix} [\mathbf{A}] & [\mathbf{BA}] \\ -[\mathbf{I}] & [\mathbf{O}] \end{vmatrix} = \det [\mathbf{BA}] .$$

7.6 **Rank of a matrix:** For an m by n matrix

(1)
$$\begin{pmatrix} a_{11} & \cdots & a_{1n} \\ \cdots & & \cdots \\ a_{m1} & \cdots & a_{mn} \end{pmatrix} ,$$

any determinant of the form

$$\begin{vmatrix} a_{i_1j_1} & \cdots & a_{i_1j_k} \\ \cdots & & \cdots \\ a_{i_kj_1} & \cdots & a_{i_kj_k} \end{vmatrix}$$

in which $i_1 < i_2 < \ldots < i_k$, and $j_1 < \ldots < j_k$, $k \leq \min. (m,n)$, is called a subdeterminant, or *minor*, of order k, of the matrix (1). If there is a positive integer r with the property that there exists a non-zero subdeterminant of (1) of order r, while every subdeterminant of order $r+1$ is zero, this r is called the *rank* of the matrix (1). The zero matrix is said to have rank zero. For any other matrix we prove that the rank is equal to the maximum number of linearly independent row vectors in the matrix. Let us assume that there exist $s \leq m$ linearly independent row vectors, and for simplicity let us assume that they are the first s rows. Clearly $s \leq n$. Let k be the least positive integer such that every subdeterminant of order k is zero. If k were less than or equal to s, by rearranging rows and columns, since there is by assumption a $(k-1)$-by-$(k-1)$ determinant which is not zero, we could have

$$(2) \qquad \begin{vmatrix} a_{11} & \cdots & a_{1\,k-1} \\ \cdots & & \cdots \\ a_{k-1,1} & \cdots & a_{k-1,k-1} \end{vmatrix} \neq 0 ,$$

while for each positive integer t, $k \leq t \leq s$

$$(3) \qquad \begin{vmatrix} a_{11} & \cdots & a_{1,k-1} & a_{1l} \\ \cdots & & \cdots & \\ a_{k-1,1} & \cdots & a_{k-1,k-1} & a_{k-1,l} \\ a_{t1} & \cdots & a_{t,k-1} & a_{tl} \end{vmatrix} = 0$$

for all positive integers $l \geq k$. Let c_i be the cofactor of a_{il} in (3). Then $c_1 a_{1l} + \cdots + c_t a_{tl} = 0$ for all $l \geq k$, and c_t (the determinant (2)) $\neq 0$. But by 3.2 (5), $c_1 a_{1j} + \cdots + c_t a_{tj} = 0$ for $j < k$, hence for all j. But the k row vectors of this matrix are linearly independent, hence every $c_i = 0$, which contradicts (2). Thus there is a subdeterminant of order s which is not zero.

Furthermore if every set of $s + 1$ row vectors is linearly dependent, every determinant of order $s + 1$ contains a row which is a linear combination of the others, hence every such determinant is zero.

Conversely, if some s-by-s determinant of the matrix (1) is not zero, the corresponding s rows of the matrix are linearly independent, while if every $(s+1)$-by-$(s+1)$ determinant is zero, every $s + 1$ rows of the matrix are linearly dependent.

Since the rank is defined in terms of determinants, clearly the rank of any matrix is equal to the rank of its transpose, and is also equal to the maximum number of linearly independent column vectors in the matrix.

7.7 **Systems of linear equations:** We treat now the problem of finding what solutions, if any, exist for a system of m equations in n unknowns.

$$(1) \qquad \begin{cases} a_{11}x_1 + \cdots + a_{1n}x_n = b_1 \\ a_{21}x_1 + \cdots + a_{2n}x_n = b_2 \\ \cdots \qquad \cdots \\ a_{m1}x_1 + \cdots + a_{mn}x_n = b_m . \end{cases}$$

Let us consider the matrices

$$[\mathbf{A}] = \begin{pmatrix} a_{11} & \cdots & a_{1n} \\ \cdots & & \cdots \\ a_{m1} & \cdots & a_{mn} \end{pmatrix} ,$$

called the *coefficient matrix* of the system (1), and

$$\begin{pmatrix} a_{11} & \dots & a_{1n} & b_1 \\ \dots & & \dots & \\ a_{m1} & \dots & a_{mn} & b_m \end{pmatrix},$$

called the *augmented matrix* of the system.

Let the rank of the coefficient matrix be r. Clearly the augmented matrix has rank either r or $r+1$. We prove in what follows that the system (1) has solutions if and only if the rank of the augmented matrix is r. If $r=0$, then every $a_{ij}=0$, hence solutions will exist, and in fact any set of numbers $x_1, \dots, x_n$ will be a solution, if and only if every $b_i=0$, so that the augmented matrix is also of rank zero. Thus we assume $r \geq 1$.

We examine first the case in which $m = n$. In this case $[\mathbf{A}]$ is a square matrix, hence has a determinant denoted by $|\mathbf{A}|$. We prove that if $|\mathbf{A}| \neq 0$, the system (1) has a unique solution. As usual, we will denote by $\mathbf{A}_{ij}$, the cofactor of a_{ij} in the determinant $|\mathbf{A}|$. We multiply the equations of (1) successively by $\mathbf{A}_{1i}, \dots,$ $\mathbf{A}_{ni}$, and add the resulting equations. By 3.2 and 7.5, the sum of the coefficients of x_j will be zero for $j \neq i$, while for $j = i$, we get

$$|\mathbf{A}|x_i = b_1\mathbf{A}_{1i} + \dots b_n\mathbf{A}_{ni} \,.$$

If we consider the determinant $|\mathbf{A}_i|$ obtained by replacing the i-th column of $|\mathbf{A}|$ by the column of b's, we observe, expanding in terms of that column, that

$$|\mathbf{A}_i| = b_1\mathbf{A}_{1i} + \dots + b_n\mathbf{A}_{ni} \,.$$

Thus we have

$$|\mathbf{A}|\, x_i = |\mathbf{A}_i| \,,$$

$$x_i = \frac{|\mathbf{A}_i|}{|\mathbf{A}|} \,.$$

We now need only show that this set $\{x_1, \dots, x_n\}$ is a solution for the original system. Then substituting in the j-th equation, we get

$$a_{j1}\frac{|\mathbf{A}_1|}{|\mathbf{A}|} + a_{j2}\frac{|\mathbf{A}_2|}{|\mathbf{A}|} + \dots + a_{jn}\frac{|\mathbf{A}_n|}{|\mathbf{A}|} = \frac{1}{|\mathbf{A}|}[a_{j1}(b_1\mathbf{A}_{11} + \dots + b_n\mathbf{A}_{n1}) + \dots + a_{jn}(b_1\mathbf{A}_{1n} + \dots + b_n\mathbf{A}_{nn})]$$

$$= \frac{1}{|\mathbf{A}|}[b_1(a_{j1}\mathbf{A}_{11} + \dots + a_{jn}\mathbf{A}_{1n}) + \dots + b_n(a_{j1}\mathbf{A}_{n1} + \dots + a_{jn}\mathbf{A}_{nn})] \,.$$

But we observe that the coefficient of b_i in the above expression is the sum of the products of the terms in the j-th row of the determinant $|\mathbf{A}|$ by the cofactors of the terms of the i-th row. Thus each coefficient is zero except the coefficient of b_j, which is $|\mathbf{A}|$. Thus

$$a_{j1}\frac{|\mathbf{A}_1|}{|\mathbf{A}|} + \dots + a_{jn}\frac{|\mathbf{A}_n|}{|\mathbf{A}|} = \frac{1}{|\mathbf{A}|}|\mathbf{A}|\, b_j = b_j \,,$$

and the x_i's above satisfy the equations of (1), the solution being unique. This rule for finding the solutions for the case $m=n$, $|\mathbf{A}| \neq 0$, is called *Cramer's Rule.*

Now we consider the general case of m equations in n unknowns, where the rank of the coefficient matrix is r. Clearly the rank of the augmented matrix is at least r. We prove first that if the system (1) has solutions, then the rank of the augmented matrix is exactly r. We may for convenience rearrange the equations of (1) if necessary so that the first r row vectors of the coefficient matrix are linearly independent. Assuming that $\{x_1, \dots, x_n\}$ is a solution of the system (1), we have for $i = 1, \dots, m$,

$$b_i = a_{i1}x_1 + \dots + a_{in}x_n \,,$$

so that the augmented matrix becomes

$$\begin{pmatrix} a_{11} & \dots & a_{1n} & a_{11}x_1 + \dots + a_{1n}x_n \\ a_{21} & \dots & a_{2n} & a_{21}x_1 + \dots + a_{2n}x_n \\ & \dots & & \dots \\ a_{m1} & \dots & a_{mn} & a_{m1}x_1 + \dots + a_{mn}x_n \end{pmatrix} .$$

Since the first r rows of the coefficient matrix are linearly independent, if m is greater than r, there exists, for every positive integer t, $r < t \leq m$, a set of numbers c_{ti} such that

$$a_{tj} = c_{t1}a_{1j} + \dots + c_{tr}a_{rj},\ j = 1, \dots, n .$$

Then since b_t is a linear combination of the numbers a_{tj},

$$b_t = c_{t1}b_1 + c_{t2}b_2 + \dots + c_{tr}b_r ,$$

that is, the term in the last column of the augmented matrix is the same linear combination of the terms $b_1, \dots b_r$, so that the t-th row of the augmented matrix is a linear combination of the first r rows. Thus since every row vector of the augmented matrix is a linear combination of the first r row vectors, any $r + 1$ row vectors of the augmented matrix are linearly dependent [see 6.6 and 7.6] and the rank of this matrix is not greater than r, hence is equal to r.

Conversely, if the augmented matrix is of rank r we show that the system (1) has solutions. Assuming, as before, that the first r rows of this matrix are linearly independent, we may rearrange the columns so that the first r columns are linearly independent column vectors of the coefficient matrix, so that

$$\begin{vmatrix} a_{11} & \dots & a_{1r} \\ \dots & & \dots \\ a_{r1} & \dots & a_{rr} \end{vmatrix} \neq 0 .$$

We write the first r equations in the form

$$\begin{cases} a_{11}x_1 + \dots + a_{1r}x_r = b_1 - a_{1,r+1}x_{r+1} - \dots - a_{1n}x_n \\ \dots \qquad \dots \qquad \dots \\ a_{r1}x_1 + \dots + a_{rr}x_r = b_r - a_{r,r+1}x_{r+1} - \dots - a_{rn}x_n \end{cases} .$$

Choose any values for $x_{r+1}, \dots, x_n$, and consider the above as a system of r equations for the r unknowns $x_1, \dots, x_r$. Since the determinant of the coefficient matrix of this system is not zero, the system has a unique solution for $x_1, \dots, x_r$. Then the set $\{x_1, \dots, x_r, x_{r+1}, \dots x_n\}$ is a solution of the first r equations of the system (1). But since any other row vector of the augmented matrix is a linear combination of these first r rows, any solution of the first r equations is a solution of the entire system.

Thus solutions of the system (1) exist if and only if the augmented matrix and the coefficient matrix have the same rank.

We observe that if we have n *homogeneous* equations ($b_1 = \dots b_n = 0$) in n unknowns, the rank of the augmented matrix is the same as the rank of the coefficient matrix. If this rank is n, the only solution of the system would be $x_1 = \dots = x_n = 0$. Non-trivial solutions will therefore exist if and only if the coefficient matrix is singular, i.e. the determinant of the coefficient matrix is zero.

7.8 **Inverse of a linear transformation:** As in 4.1, we define the inverse of a linear transformation $\mathbf{A}$ to be a transformation $\mathbf{A}^{-1}$ such that $\mathbf{A}\mathbf{A}^{-1} = \mathbf{I}$. If such a transformation exists it follows also that $\mathbf{A}^{-1}\mathbf{A} = \mathbf{I}$.

Let the matrix of **A** with respect to a given base be

$$\begin{pmatrix} a_{11} & \dots & a_{1n} \\ \dots & & \dots \\ a_{n1} & \dots & a_{nn} \end{pmatrix} .$$

The matrix of $\mathbf{A}^{-1}$ with respect to the same base is found as follows: Suppose for $\xi = (x_1, \dots, x_n)$ we have $\mathbf{A}\,\xi = (y_1, \dots, y_n)$. By 7.4 we get

$$\begin{cases} y_1 = a_{11}x_1 + \dots + a_{n1}x_n \\ \dots \quad \dots \quad \dots \\ y_n = a_{1n}x_1 + \dots + a_{nn}x_n \, . \end{cases}$$

In order to have a unique solution for $x_1, \dots, x_n$ we must have

$$\begin{vmatrix} a_{11} & \dots & a_{n1} \\ \dots & & \dots \\ a_{1n} & \dots & a_{nn} \end{vmatrix} = \begin{vmatrix} a_{11} & \dots & a_{1n} \\ \dots & & \dots \\ a_{n1} & \dots & a_{nn} \end{vmatrix} \neq 0 \, .$$

It is left to the reader to show that the matrix of $\mathbf{A}^{-1}$ is

$$\frac{1}{\det \mathbf{A}} \begin{pmatrix} \mathbf{A}_{11} & \dots & \mathbf{A}_{n1} \\ \dots & & \dots \\ \mathbf{A}_{1n} & \dots & \mathbf{A}_{nn} \end{pmatrix}, \quad \text{where } \mathbf{A}_{ij} \text{ is the cofactor of } a_{ij}.$$

A transformation is called non-singular if its inverse exists.

The theorem 4.3 is true in general and can be proved in exactly the same way.

7.9 **Adjoint of a transformation:** For any transformation **A** we define the adjoint $\mathbf{A}^*$, by the equality $(\mathbf{A}\xi, \eta) = (\xi, \mathbf{A}^*\eta)$, for all vectors ξ and η.

The matrix of $\mathbf{A}^*$ is the conjugate transpose of **A**, i.e., the transpose of the matrix whose elements are the complex conjugates of the corresponding elements of the matrix of **A**. We leave the proof to the reader.

A theorem similar to 4.6 is true, that is,

$$(\mathbf{A} + \mathbf{B})^* = \mathbf{A}^* + \mathbf{B}^*, \text{ and}$$

$$(\mathbf{AB})^* = \mathbf{B}^*\mathbf{A}^*. \text{ (Note the order).}$$

The proof is identical to the one in 4.6

As in 4.7 we can prove that if $\mathbf{A}^{-1}$ exists,

$$(\mathbf{A}^*)^{-1} = (\mathbf{A}^{-1})^* \, .$$

7.10 **Unitary Transformation:** A *unitary* transformation is also defined, as in 4.8, as one which preserves inner product.

We shall prove that the rows of the matrix of a unitary transformation with respect to an orthonormal base form an orthonormal set of vectors. Let **A** be unitary and $\{\alpha_1, \dots, \alpha_n\}$ an orthonormal base.

Then
$$\mathbf{A}\,\alpha_1 = a_{11}\,\alpha_1 + \dots + a_{1n}\,\alpha_n \, ,$$
$$\dots \quad \dots \quad \dots$$
$$\mathbf{A}\alpha_n = a_{n1}\,\alpha_1 + \dots + a_{nn}\,\alpha_n \, . \quad \text{But}$$
$$1 = (\alpha_i, \alpha_i) = (\mathbf{A}\,\alpha_i, \mathbf{A}\,\alpha_i) = |\mathbf{A}\,\alpha_i|^2 = |a_{i1}|^2 + \dots + |a_{in}|^2 \, ,$$

for $i = 1, \ldots, n$, and

$$0 = (\alpha_i, \alpha_j) = (\mathbf{A}\alpha_i, \mathbf{A}\alpha_j) = a_{i1}\bar{a}_{j1} + \ldots + a_{in}\bar{a}_{jn}, \; i \neq j \text{ and } i, j = 1, \ldots, n.$$

This clearly proves the statement. The same theorem is true for columns.

It is also true that $\mathbf{A}^{-1} = \mathbf{A}^*$ if and only if $\mathbf{A}$ is unitary. The proof is similar to the one in 4.9.

7.11 **Change of Base:** Let $\{\alpha_1, \ldots, \alpha_n\}$ be a base for V_n. Let $\{\beta_1, \ldots, \beta_n\}$ be another base. Let $\mathbf{A}$ be a transformation on V_n whose matrix with respect to $\{\alpha_1, \ldots, \alpha_n\}$ is

$$[\mathbf{A}] = \begin{pmatrix} a_{11} & \ldots & a_{1n} \\ \ldots & & \ldots \\ a_{n1} & \ldots & a_{nn} \end{pmatrix}.$$

If the matrix of $\mathbf{A}$ with respect to $\{\beta_1, \ldots, \beta_n\}$ is

$$[\mathbf{B}] = \begin{pmatrix} b_{11} & \ldots & b_{1n} \\ \ldots & & \ldots \\ b_{n1} & \ldots & b_{nn} \end{pmatrix},$$

then $[\mathbf{B}] = [\mathbf{D}]\,[\mathbf{A}]\,[\mathbf{D}^{-1}]$, where $[\mathbf{D}]$ is the matrix of the change of base from $\{\alpha_1, \ldots, \alpha_n\}$ to $\{\beta_1, \ldots, \beta_n\}$. The proof is exactly like the one in 4.10.

Note that if $\{\alpha_1, \ldots, \alpha_n\}$ and $\{\beta_1, \ldots, \beta_n\}$ are orthonormal, then $\mathbf{D}$ is unitary, and the matrix of $\mathbf{A}$ after the change of base will be $[\mathbf{B}] = [\mathbf{D}]\,[\mathbf{A}]\,[\mathbf{D}^*]$.

7.12 **Characteristic values and Characteristic vectors of a transformation:** For a transformation $\mathbf{A}$ on V_n we define characteristic values and characteristic vectors as in 5.1. Clearly in order that

$$(\mathbf{A} - \lambda\,\mathbf{I})\,\xi = 0, \text{ or}$$

$$\begin{cases} (a_{11} - \lambda)\,x_1 + a_{21}x_2 + \ldots + a_{n1}x_n = 0 \\ \ldots \qquad \ldots \qquad \ldots \\ a_{1n}x_1 + a_{2n}x_2 + \ldots + (a_{nn} - \lambda)\,x_n = 0\,. \end{cases}$$

have non-zero solutions we must have

$$(1) \qquad \begin{vmatrix} a_{11}-\lambda & \ldots & a_{1n} \\ \ldots & \ldots & \ldots \\ a_{n1} & \ldots & a_{nn}-\lambda \end{vmatrix} = |\mathbf{A}-\lambda\,\mathbf{I}| = 0$$

[see the last paragraph of 7.7].

We call (1) the characteristic equation of $\mathbf{A}$ and its roots the characteristic values of $\mathbf{A}$. As was shown in 5.2 the characteristic equation and consequently the characteristic values and characteristic vectors of a transformation are independent of the choice of base. The same proof carries over.

It is also clear that the trace, i.e., $a_{11} + \ldots + a_{nn}$, and the determinant of a matrix are respectively the sum and the product of its characteristic values. Theorem 5.5 is also independent of the dimension of the space.

7.13 **Definition:** A transformation $\mathbf{A}$ on E_n is called *normal* if $\mathbf{AA}^* = \mathbf{A}^*\mathbf{A}$. A transformation $\mathbf{A}$ on E_n is called *Hermitian* if $\mathbf{A}^* = \mathbf{A}$.

7.14 **Theorem:** The characteristic values of a Hermitian transformation **A** are real. Moreover the characteristic vectors corresponding to different characteristic values are orthogonal.

Proof: Let ξ be a non-zero characteristic vector of E_n corresponding to λ, a characteristic value of **A**. Then

$$\lambda(\xi,\xi) = (\mathbf{A}\xi,\xi) = (\xi,\mathbf{A}^*\xi) = (\xi,\mathbf{A}\xi) = (\xi,\lambda\xi) = \bar{\lambda}(\xi,\xi) .$$

Since $\xi \neq 0$ we have $\lambda = \bar{\lambda}$ and this proves the first part of the theorem. Suppose $\lambda_1 \neq \lambda_2$ and $\mathbf{A}\xi = \lambda_1\xi$, $\mathbf{A}\eta = \lambda_2\eta$, where ξ and η are non-zero vectors. Then

$$\lambda_1(\xi,\eta) = (\mathbf{A}\xi,\eta) = (\xi,\mathbf{A}^*\eta) = (\xi,\mathbf{A}\eta) = \lambda_2(\xi,\eta) .$$

Therefore $(\lambda_1 - \lambda_2)(\xi,\eta) = 0$

Since $\lambda_1 \neq \lambda_2$ we have $(\xi,\eta) = 0$ and this proves the second part of the theorem.

7.15 **Theorem:** There is always a base for E_n with respect to which the matrix of a Hermitian transformation has diagonal form, i.e., we have zeros everywhere except possibly on the main diagonal.

Proof: Let λ_1 be a characteristic value of **A**. As in 5.10 let α_1 be a characteristic vector of unit norm corresponding to λ_1. If we choose an orthonormal base $\{\alpha_1, \beta_2, \ldots, \beta_n\}$ whose first element is α_1, then $\mathbf{A}\alpha_1 = \lambda_1\alpha_1 = (\lambda_1, 0, \ldots, 0)$ and the matrix of **A** with respect to this base will be of the form

$$\begin{pmatrix} \lambda_1 & 0 \ldots 0 \\ 0 & \\ \cdot & \\ \cdot & (\mathbf{B}) \\ \cdot & \\ 0 & \end{pmatrix}$$

If all λ's are distinct this clearly proves the theorem.

Now suppose $\lambda_1, \lambda_2, \ldots, \lambda_k$ are k distinct characteristic values of **A**, with multiplicities respectively $h_1, \ldots, h_k$. Let $\alpha_1, \ldots, \alpha_k$ be characteristic vectors of unit norm corresponding to $\lambda_1, \ldots, \lambda_k$. By 7.14, $\{\alpha_1, \ldots, \alpha_k\}$ is an orthonormal set of vectors. Let $\{\alpha_1, \ldots, \alpha_k, \beta_{k+1}, \ldots, \beta_n\}$ be an orthonormal base for E_n. With respect to this base we have

$$\begin{pmatrix} \lambda_1 & 0 \ldots & \ldots & 0 \\ 0 & \lambda_2 . & & \\ \vdots & & \ddots \lambda_k & 0 \ldots \\ 0 & & 0 & [\mathbf{B}] \end{pmatrix} ,$$

where [**B**] is an $(n\text{-}k)$ – by – $(n - k)$ matrix

By 7.12 the characteristic equations of **A** will be

$$(\lambda_1 - \lambda) \cdot \ldots \cdot (\lambda_k - \lambda). \; |\mathbf{B} - \lambda\mathbf{I}| = 0 .$$

This implies that the characteristic values of [**B**] are $\lambda_1, \ldots, \lambda_k$ with the multiplicities $h_1 - 1, \ldots, h_k - 1$. Here in the case that some λ_i is a single root of $|\mathbf{A} - \lambda\mathbf{I}| = 0$ we eliminate that λ_i from $\lambda_1, \ldots, \lambda_k$. By a method similar to that of the previous paragraph we can get a base in the subspace generated by $\beta_{k+1}, \ldots, \beta_n$, with respect to which the matrix of **B** has the form.

$$\begin{pmatrix} \lambda_1 & \ldots & 0 \\ \cdot & \ddots \lambda_k & \\ \cdot & & \\ \cdot & & \\ 0 & & [\mathbf{C}] \end{pmatrix} ,$$

where [**C**] is a matrix of fewer elements than [**B**]. This intuitively shows how we can finally change the matrix of **A** to diagonal form. Since the change of base is from one orthonormal base to another, there is a unitary transformation **P** such that

$$[\mathbf{P}]\,[\mathbf{A}]\,[\mathbf{P}^*] = [\Lambda]\,, \quad \text{where}$$

$[\Lambda]$ is in diagonal form.

EXCERCISES 7

1. Find the characteristic values and characteristic vectors of:

(i) $\begin{pmatrix} 2 & 4 \\ 5 & 3 \end{pmatrix}$, (ii) $\begin{pmatrix} 4 & 9 & 0 \\ 0 & -2 & 8 \\ 0 & 0 & 7 \end{pmatrix}$.

2. Without solving give the characteristic values of

$$\begin{pmatrix} 2 & 1 & 0 \\ 0 & 1 & 1 \\ 0 & 0 & -1 \end{pmatrix}$$

3. Let $\mathbf{A}\,\xi = (\xi,\epsilon)\,\epsilon$, where ξ is a vector in E_n and ϵ is vector of unit norm.

(i) Show that **A** is a linear transformation.

(ii) Write the matrix of **A** with respect to an orthonormal base, first containing ϵ, and second not containing ϵ.

(iii) Find the characteristic values of **A**.

4. Show that

$$\begin{vmatrix} 1 & x_1 & x_1^2 & \cdots & x_1^{n-1} \\ 1 & x_2 & x_2^2 & \cdots & x_2^{n-1} \\ \cdots & & & \cdots & \\ 1 & x_n & x_n^2 & \cdots & x_n^{n-1} \end{vmatrix} = \prod_{\substack{i,j=1 \\ i<j}}^{n} (x_i - x_j) = (x_1 - x_2)\,(x_1 - x_3) \cdot \ldots \cdot (x_{n-1} - x_n).$$

5.* If each element of a square matrix [**A**] is a function of x, then show that

(i) $\mathbf{A}_{ij} = \dfrac{\partial\,(\det \mathbf{A})}{\partial\, a_{ij}}$.

(ii) $\dfrac{d}{dx}(\det \mathbf{A}) = \displaystyle\sum_{j,\,k=1}^{n} \frac{d\,a_{ik}}{dx}\,\mathbf{A}_{jk},\ i=1,\ldots,n$.

6. Let $\mathbf{A}_1 = \dfrac{d}{dx},\ \mathbf{A}_2 = \dfrac{d^2}{dx^2},\ \ldots,\ \mathbf{A}_k = \dfrac{d^k}{dx^k}$ on the vector space of chapter 6, exercise 2. Show that $\mathbf{A}_1, \ldots, \mathbf{A}_k$ are linear transformations.

7. Determine the rank of each of the following matrices.

(i) $\begin{pmatrix} 3 & 1 & 0 & 2 \\ 1 & -1 & 2 & -1 \\ 1 & 3 & -4 & 4 \end{pmatrix}$, (ii) $\begin{pmatrix} 1 & 4 & -1 & 2 & 2 \\ 3 & -2 & 1 & 1 & 0 \\ -2 & -1 & 3 & 2 & 0 \end{pmatrix}$,

(iii) $\begin{pmatrix} 3 & i & -1 & 0 \\ 1+i & 1 & -2 & 1 \\ 5+2i & 2+i & -5 & 2 \end{pmatrix}$, (iv) $\begin{pmatrix} 1 & 1+i & 3i \\ 0 & 1 & 1-i \\ 2 & 1 & 1+i \end{pmatrix}$.

8. Evaluate:

(i) $\begin{vmatrix} 1 & 3 & 0 & 1 \\ 2 & -1 & 2 & 1 \\ -3 & -2 & 1 & 0 \\ 3 & 3 & 1 & 2 \end{vmatrix}$, (ii) $\begin{vmatrix} 2 & 1+i & 3i \\ -1 & -2i & 0 \\ 1+2i & 4 & 1-2i \end{vmatrix}$.

9. Find the characteristic values of:

(i) $\begin{pmatrix} 1 & 0 & 1 & -5 \\ 0 & -2 & 2 & 1 \\ 0 & 0 & -3 & 1 \\ 0 & 0 & 0 & 2 \end{pmatrix}$, (ii) $\begin{pmatrix} \frac{7}{4} & 0 & \frac{\sqrt{3}}{4} & 0 \\ 0 & \frac{-41}{25} & 0 & \frac{-12}{25} \\ \frac{\sqrt{3}}{4} & 0 & \frac{5}{4} & 0 \\ 0 & \frac{-12}{25} & 0 & \frac{-34}{25} \end{pmatrix}$, (iii) $\begin{pmatrix} \frac{5}{4} & \frac{\sqrt{6}}{12} & \frac{3\sqrt{2}}{4} & \frac{-7\sqrt{3}}{12} \\ \frac{\sqrt{6}}{12} & \frac{5}{6} & \frac{-\sqrt{3}}{2} & \frac{\sqrt{2}}{12} \\ \frac{3\sqrt{2}}{4} & \frac{-\sqrt{3}}{2} & \frac{1}{2} & \frac{\sqrt{6}}{4} \\ \frac{-7\sqrt{3}}{12} & \frac{\sqrt{2}}{12} & \frac{\sqrt{6}}{4} & \frac{29}{12} \end{pmatrix}$.

10. Find the matrices of $\mathbf{A}_1, \ldots, \mathbf{A}_k$ of 6. with respect to the base given in chapter 6, exercise 2, and find the rank of each of these matrices.

11. Show that the equation of the circle passing through (x_1,y_1), (x_2,y_2), (x_3,y_3) is

$$\begin{vmatrix} x^2+y^2 & x & y_1 & 1 \\ x_1^2+y_1^2 & x_1 & y_1 & 1 \\ x_2^2+y_2^2 & x_2 & y_2 & 1 \\ x_3^2+y_3^2 & x_3 & y_3 & 1 \end{vmatrix} = 0$$

12. Find all solutions, if any exist, of

(i) $\begin{cases} x + y + z + = 5 \\ 2x + y - z + w = 1 \\ x + 2y - z + w = 2 \\ y + 2z + 3w = 3 \end{cases}$ (ii) $\begin{cases} x + y - z = 1 \\ x + 2y + z - w = 8 \\ 2x - y - 3w = 3 \\ 3x + 3y + 5z - 6w = 5 \end{cases}$

(iii) $\begin{cases} x + 2y - z = 3 \\ 2x - y + w = -1 \\ -x + 2y + 2z + 3w = 4 \\ 3x - 5y + 4z + 6w = 5 \end{cases}$ (iv) $\begin{cases} 2x + y - z + w = 1 \\ 3x - 2y + z - 3w = 4 \\ x + 4y - 3z + 5w = -2 \end{cases}$.

13. Change to diagonal form

(i) $\begin{pmatrix} 3 & 1+i \\ 1-i & 2 \end{pmatrix}$ (ii) $\begin{pmatrix} \frac{7}{4} & \frac{-3i\sqrt{3}}{4} & 0 \\ \frac{3i\sqrt{3}}{4} & \frac{5}{4} & 0 \\ 0 & 0 & 3 \end{pmatrix}$.

14. Consider the following subspaces of the space of all polynomials of degree two. [see chapter 6, exercise 2].

(1) the space generated by $\{x^2, x\}$
(2) the space generated by $\{x, 1\}$

(i) Show that $\mathbf{A} = \frac{d}{dx}$ transforms (1) onto (2) and $\mathbf{B} = \int_0^x (\ldots)\, dx$ transforms (2) onto (1).

(ii) Find the matrices of **A** and **B** with respect to the whole space and compare.

15. Change to diagonal form

(i) $\begin{pmatrix} 1 & 0 & -1 \\ 0 & 7 & 0 \\ -1 & 0 & 2 \end{pmatrix}$, (ii) $\begin{pmatrix} 0 & 1 & 0 \\ 1 & 0 & 2 \\ 0 & 2 & 0 \end{pmatrix}$.

8. QUADRATIC FORMS AND APPLICATION TO GEOMETRY

8.1 **Definition:** In this chapter scalars are real numbers. In a unitary space E_n we define a *quadratic form* to be

(1) $$Q = (x_1\ x_2 \dots x_n) \begin{pmatrix} a_{11} & b_2 & b_3 & \dots & b_n \\ b_2 & a_{22} & c_3 & \dots & c_n \\ b_3 & c_3 & & \dots & \\ \dots & & & & \\ b_n & c_n & & \dots & a_{nn} \end{pmatrix} \begin{pmatrix} x_1 \\ x_2 \\ \cdot \\ \cdot \\ \cdot \\ x_n \end{pmatrix}.$$

We can write (1) as $Q = (\mathbf{A}\xi, \xi)$ where $\mathbf{A}$ is Hermitian and $\xi = (x_1, \dots, x_n)$. It is clear that any (homogeneous) quadratic form can be written in the form (1). For an example look at 5.11.

8.2 **Reduction of a quadratic form to canonical form:** The technique is the same as 5.11. We shall carry out the process again.

(1) $$Q = (x_1 \dots x_n)\ [\mathbf{A}] \begin{pmatrix} x_1 \\ \cdot \\ \cdot \\ \cdot \\ x_n \end{pmatrix}$$

We can write (1) as follows:

$$Q = (x \dots x_n)\ [\mathbf{I}]\ [\mathbf{A}]\ [\mathbf{I}] \begin{pmatrix} x_1 \\ \cdot \\ \cdot \\ \cdot \\ x_n \end{pmatrix} \quad \text{or}$$

$$Q = (x_1 \dots x_n)\ [\mathbf{U}]\ [\mathbf{U}']\ [\mathbf{A}]\ [\mathbf{U}]\ [\mathbf{U}'] \begin{pmatrix} x_1 \\ \cdot \\ \cdot \\ \cdot \\ x_n \end{pmatrix}, \quad \text{where } \mathbf{U} \text{ is unitary.}$$

But $\mathbf{U}$ can be chosen such that $[\mathbf{U}']\ [\mathbf{A}]\ [\mathbf{U}]$ is of diagonal form, say $\begin{pmatrix} \lambda_1 & & 0 \\ & \ddots & \\ 0 & & \lambda_n \end{pmatrix}$. Let $(x_1 \dots x_n)\ [\mathbf{U}] = (y_1 \dots y_n)$ and clearly $[\mathbf{U}'] \begin{pmatrix} x_1 \\ \vdots \\ x_n \end{pmatrix} = \begin{pmatrix} y_1 \\ \vdots \\ y_n \end{pmatrix}$. Therefore

(2) $$Q = (y_1 \dots y_n) \begin{pmatrix} \lambda_1 & & 0 \\ & \ddots & \\ 0 & & \lambda_n \end{pmatrix} \begin{pmatrix} y_1 \\ \vdots \\ y_n \end{pmatrix}$$

$$= \lambda_1 y_1^2 + \dots + \lambda_n y_n^2 .$$

If all λ's are non-negative (positive) we say Q is non-negative (positive).

8.3 **Reduction to Sum or difference of squares:** If we let $y_i \sqrt{|\lambda_i|} = z_i$, $\lambda_i \neq 0$, $i = 1, \ldots, k \leq n$, then 8.2 (2) will be $Q = \pm z_1^2 \pm \ldots \pm z_k^2$. The sign is taken the same as the sign of the corresponding characteristic value. Note that this transformation is not unitary.

8.4 **Simultaneous reduction of two quadratic forms:**

Let $Q_1 = (x_1 \ldots x_n)\ [\mathbf{A}] \begin{pmatrix} x_1 \\ \cdot \\ \cdot \\ \cdot \\ x_n \end{pmatrix}$ and

$Q_2 = (y_1 \ldots y_n)\ [\mathbf{B}] \begin{pmatrix} y_1 \\ \cdot \\ \cdot \\ \cdot \\ y_n \end{pmatrix}$ be two quadratic forms where

Q_1 is positive. We can reduce Q_1 to the form

(1) $\quad Q_1 = x_1^2 + \ldots + x_n^2 \quad$ [by 8.3] .

Clearly any unitary transformation does not change (1). Let Q_2 be changed to canonical form $Q_2 = \mu_1 y_1^2 + \ldots + \mu_n y_n^2$. But (1) implies that there is a change of base **S** such that $[\mathbf{S}]\ [\mathbf{A}]\ [\mathbf{S}'] = [\mathbf{I}]$ and this change of base changes $[\mathbf{B}]$ to $[\mathbf{S}]\ [\mathbf{B}]\ [\mathbf{S}'] = [\mathbf{B}_1]$. Therefore

$$0 = |\mathbf{B}_1 - \mu \mathbf{I}| = |\mathbf{S\ B\ S}' - \mu \mathbf{I}| = |\mathbf{S\ B\ S}' - \mu \mathbf{S\ A\ S}'| = |\mathbf{S}|\ |\mathbf{B} - \mu \mathbf{A}|\ |\mathbf{S}'|, \text{ or}$$

(2) $\quad |\mathbf{B} - \mu \mathbf{A}| = 0$. That is, the μ's are the roots of (2).

8.5 **Homogeneous Coordinates:** It is convenient for some purposes to designate a vector ξ in an n-dimensional space by an ordered set of $n + 1$ scalars, $(x_1, \ldots, x_{n+1})$ with $x_{n+1} \neq 0$, where the above set denotes the vector

(1) $$\xi = \left(\frac{x_1}{x_{n+1}}, \ldots, \frac{x_n}{x_{n+1}}\right) .$$

The two sets $(x_1, \ldots, x_{n+1})$ and $(y_1, \ldots, y_{n+1})$ yield the same vector if there is a scalar $k \neq 0$ such that $y_i = k\ x_i$, $i = 1, \ldots, n+1$. A set of scalars $(x_1, \ldots, x_{n+1})$ satisfying (1) is called a set of *homogeneous coordinates* of ξ.

In what follows we are concerned with the spaces of two or three dimensions. We observe that if we introduce homogeneous coordinates into polynomial equations in two or three variables, the equations become homogeneous in three or four variables respectively. In particular, equations of second degree may be written as quadratic forms in homogeneous coordinates.

8.6 **Change of coordinate system:** We recall from the study of analytic geometry that a transformation of coordinates in the Euclidean plane was a combination of a translation,

$$\begin{cases} x = x' + x_1 \\ y = y' + y_1 \end{cases}$$

and a rotation about the origin (a change from one orthonormal base to another)

$$\begin{cases} x' = x'' \cos\theta - y'' \sin\theta \\ y' = x'' \sin\theta + y'' \cos\theta \ . \end{cases}$$

Thus any rigid motion in the plane (Euclidean transformation) can be written as

$$\begin{cases} x = x' \cos\theta - y' \sin\theta + x_1 \\ y = x' \sin\theta + y' \cos\theta + y_1 \,. \end{cases}$$

Note that this transformation is not linear [see 2.1]. If we introduce homogeneous coordinates, letting x, y, w be the coordinates of a point (vector) and x', y', w' be the coordinates of the same point in the new base, then the general Euclidean transformation takes the form of the linear transformation,

$$\begin{cases} x = x' \cos\theta - y' \sin\theta + w'x_1 \\ y = x' \sin\theta + y' \cos\theta + w'y_1 \\ w = \qquad\qquad\qquad\qquad w' \,, \end{cases}$$

and the transpose of the matrix of this transformation is

$$\begin{pmatrix} \cos\theta & -\sin\theta & x_1 \\ \sin\theta & \cos\theta & y_1 \\ 0 & 0 & 1 \end{pmatrix},$$

which has determinant equal to 1, while the matrix obtained from it by removing its last row and column is the matrix of the rotation of the coordinate system about the origin, and is therefore unitary.

In similar fashion, it can be shown that in a Euclidean three-dimensional space a rigid motion (change of coordinate system) is defined as a linear transformation.

$$\begin{cases} x = a_{11} x' + a_{12} y' + a_{13} z' + x_1 w' \\ y = a_{21} x' + a_{22} y' + a_{23} z' + y_1 w' \\ z = a_{31} x' + a_{32} y' + a_{33} z' + z_1 w' \\ w = \qquad\qquad\qquad\qquad\qquad w' \,, \end{cases}$$

where (x, y, z, w) and (x', y', z', w') are the homogeneous coordinates of the same point with respect to the two coordinate systems, and the matrix.

$$\begin{pmatrix} a_{11} & a_{12} & a_{13} \\ a_{21} & a_{22} & a_{23} \\ a_{31} & a_{32} & a_{33} \end{pmatrix}$$

is unitary, while the matrix of the transformation itself has determinant equal to 1.

8.7 **Invariance of rank:** In what follows we will have to make use of the fact that if **B** is any matrix, and **A** is a non-singular square matrix, and if **AB** is defined, then the rank of **AB** is equal to the rank of **B**. To prove this we need only recall that the rank of **B** is equal to the maximum number of linearly independent column vectors in **B** [see 7.6]. The column vectors of the matrix **AB** are the vectors obtained by applying the transformation with matrix **A**' to the column vectors of **B** [see 7.3]. If **A** is non-singular, $\mathbf{A}^{-1}$ and hence $(\mathbf{A}')^{-1}$ exist [see 7.9], whence any linear relation (dependence or independence) among the column vectors of **B** gives the same relation among those of **AB**, and any relation among the column vectors of **AB** gives the same relation among those of **B**. Thus **B** and **AB** have the same rank.

Similarly, if **A** is non-singular and **BA** is defined, then **B** and **BA** have the same rank. We thus observe that if **A** and **B** are square matrices of the same size, and if $\mathbf{A}^{-1}$ exists, then $\mathbf{A}'\mathbf{B}\mathbf{A}$, $\mathbf{A}\mathbf{B}\mathbf{A}'$, $\mathbf{A}^{-1}\mathbf{B}\mathbf{A}$, and $\mathbf{A}\mathbf{B}\mathbf{A}^{-1}$ have the same rank as **B**.

Further, if **B** is symmetric, since there exists a non-singular matrix **A** such that $\mathbf{ABA}^{-1}$ is in diagonal form [see 7.5], the rank of **B** is equal to the number of non-zero characteristic roots of **B**.

8.8 **Second degree curves:** Let us consider the general equation of second degree in two variables, with real coefficients, written in the form

$$(1) \qquad ax^2 + 2\,h\,xy + b\,y^2 + 2\,p\,x + 2\,q\,y + d = 0$$

If we introduce homogeneous coordinates in this equation, writing $\frac{x}{w}$ for x and $\frac{y}{w}$ for y, we get

$$(2) \qquad ax^2 + 2\,h\,xy + b\,y^2 + 2\,p\,x\,w + 2\,q\,y\,w + d\,w^2 = 0\,.$$

Equation (2) is a quadratic form, from which (1) may be obtained by setting $w = 1$. We write (2) in the form

$$(3) \qquad (x\;y\;w)\begin{pmatrix} a & h & p \\ h & b & q \\ p & q & d \end{pmatrix}\begin{pmatrix} x \\ y \\ w \end{pmatrix} = (0)\,.$$

Let us determine the effect of a change of coordinates on equation (3). We write

$$\mathbf{V} = \begin{pmatrix} x \\ y \\ w \end{pmatrix}, \quad \mathbf{V}_1 = \begin{pmatrix} x' \\ y' \\ w' \end{pmatrix}, \quad \mathbf{S} = \begin{pmatrix} \cos\theta & -\sin\theta & x_1 \\ \sin\theta & \cos\theta & y_1 \\ 0 & 0 & 1 \end{pmatrix}, \quad \mathbf{A} = \begin{pmatrix} a & h & p \\ h & b & q \\ p & q & d \end{pmatrix}$$

so that $\mathbf{V} = \mathbf{SV}_1$, whence, by 7.9 $\mathbf{V}' = \mathbf{V}_1'\,\mathbf{S}'$. Thus we may write (3) as $\mathbf{V}'\mathbf{AV} = \mathbf{O}$, and the transformed equation, also a quadratic form, as $\mathbf{V}_1'\mathbf{BV}_1 = \mathbf{O}$, where

$$\mathbf{V}'\mathbf{AV} = \mathbf{V}_1'\mathbf{S}'\,\mathbf{A}\,\mathbf{S}\,\mathbf{V}_1 = \mathbf{V}_1'\,\mathbf{B}\,\mathbf{V}_1\,,$$

so that $\qquad \mathbf{B} = \mathbf{S}'\,\mathbf{A}\,\mathbf{S}\,.$

If we now write $\qquad \mathbf{Q} = \begin{pmatrix} a & h \\ h & b \end{pmatrix}, \quad \mathbf{T} = \begin{pmatrix} \cos\theta & -\sin\theta \\ \sin\theta & \cos\theta \end{pmatrix},$

and denote by **R** the two-by-two matrix obtained in like fashion from **B**, we observe that **R** is then independent of x_1 and y_1, and

$$\mathbf{R} = \mathbf{T}'\,\mathbf{Q}\,\mathbf{T} = \mathbf{T}^{-1}\,\mathbf{Q}\,\mathbf{T}\,,$$

since **T** is unitary. The rank, determinant, and characteristic equation of **R** are the same as those of **Q** [see 7.12, 8.7], while the rank and determinant of **B** are the same as those of **A**, since the determinant of **S** is equal to 1. Finally, since **Q** is symmetric, **T** may be chosen so that **R** is in diagonal form, [see 7.15], so that equation (2) becomes

$$(2') \qquad a'x'^2 + b'y'^2 + 2p'x'w' + 2q'y'w' + d'w'^2 = 0\,,$$

whence (1) becomes

$$(1') \qquad a'x'^2 + b'y'^2 + 2p'x' + 2q'y' + d' = 0$$

Let us now consider the standard form of the equations of second degree in two variables. These equations may be written and identified as follows:

i $\qquad \dfrac{x^2}{a^2} + \dfrac{y^2}{b^2} - 1 = 0,$ the real ellipse,

ii $\qquad \dfrac{x^2}{a^2} + \dfrac{y^2}{b^2} + 1 = 0,$ the imaginary ellipse, iii $\frac{x^2}{a^2} + \frac{y^2}{b^2} = 0$, imaginary intersecting straight lines,

iv $\frac{x^2}{a^2} - \frac{y^2}{b^2} + 1 = 0$, the hyperbola,

v $\frac{x^2}{a^2} - \frac{y^2}{b^2} = 0$, real intersecting straight lines,

vi $x^2 + 2py = 0$, the parabola,

vii $x^2 + a^2 = 0$, imaginary parallel lines,

viii $x^2 - a^2 = 0$, real parallel lines,

ix $x^2 = 0$, two coincident lines.

We observe that the matrix **B** is of rank 3 for the ellipse, hyperbola, and parabola, of rank 2 for pairs of distinct lines, and of rank 1 for the case of coincident lines. For the ellipse, the product of the characteristic values of the matrix **R** is positive, for the hyperbola it is negative and for the parabola it is zero (**R** is of rank 1). But the product of the characteristic values of **R** is equal to the determinant of **R**, which is equal to the determinant of **Q**, that is

$$|\mathbf{Q}| = \lambda_1 \lambda_2 = ab - h^2 .$$

Thus an irreducible second degree curve is a hyperbola if $h^2 - ab > 0$, an ellipse if $h^2 - ab < 0$, and a parabola if $h^2 - ab = 0$. Let us consider explicitly the matrix $\mathbf{B} = \mathbf{S'AS}$, that is

$$\begin{pmatrix} \cos\theta & \sin\theta & 0 \\ -\sin\theta & \cos\theta & 0 \\ x_1 & y_1 & 1 \end{pmatrix} \begin{pmatrix} a & h & p \\ h & b & q \\ p & q & d \end{pmatrix} \begin{pmatrix} \cos\theta & -\sin\theta & x_1 \\ \sin\theta & \cos\theta & y_1 \\ 0 & 0 & 1 \end{pmatrix} .$$

Then if **B** is the matrix

$$\begin{pmatrix} a' & h' & p' \\ h' & b' & q' \\ p' & q' & d' \end{pmatrix} ,$$

we have

$$a' = a\cos^2\theta + 2h\sin\theta\cos\theta + b\sin^2\theta ,$$

$$h' = (b-a)\sin\theta\cos\theta + h(\cos^2\theta - \sin^2\theta) ,$$

$$b' = a\sin^2\theta - 2h\sin\theta\cos\theta + b\cos^2\theta ,$$

and θ may be chosen so that $h' = 0$. Further,

$$p' = (a\cos\theta + h\sin\theta)x_1 + (h\cos\theta + b\sin\theta)y_1 + p\cos\theta + q\sin\theta ,$$

$$q' = (-a\sin\theta + h\cos\theta)x_1 + (-h\sin\theta + b\cos\theta)y_1 - p\sin\theta + q\cos\theta ,$$

$$d' = ax_1^2 + 2hx_1y_1 + by_1^2 + 2px_1 + 2qy_1 + d .$$

We observe that it will be possible to choose x_1 and y_1 so that p' and q' both become zero if the determinant of their coefficients is not zero. But this gives

$$\begin{vmatrix} a\cos\theta + h\sin\theta & h\cos\theta + b\sin\theta \\ -a\sin\theta + h\cos\theta & -h\sin\theta + b\cos\theta \end{vmatrix} = ab - h^2 \neq 0 .$$

Thus if the rank of **Q** is 2, this unique choice will give a value of d' which then reduces equation (1) to one of the forms $i - v$. Otherwise, if $ab - h^2 = 0$, it can be shown that $a + b \neq 0$, and that the angle θ which will make $h' = 0$ can be chosen such that

$$\cos\theta = \sqrt{\frac{a}{a+b}}, \quad \sin\theta = \pm\sqrt{\frac{b}{a+b}}, \quad \text{the sign being the sign of } ah.$$

This will also give $a' = a + b$, $b' = 0$ and

$$p' = x_1\sqrt{a(a+b)} \pm y_1\sqrt{b(a+b)} + p\sqrt{\frac{a}{a+b}} \pm q\sqrt{\frac{b}{a+b}},$$

$$q' = q\sqrt{\frac{a}{a+b}} - \left(\pm p\sqrt{\frac{b}{a+b}}\right).$$

Then if $q' \neq 0$, $d' = \dfrac{1}{a+b}\,p'^2 + lx_1 + my_1 + n$, and it can be shown that x_1 and y_1 may be chosen so that p' and d' are zero and the equation (1') reduces to the form vi.

Finally, if $q' = 0$, $d' = \dfrac{1}{a+b}\,p'^2 + r$, and we may choose x_1 and y_1 so that $p' = 0$. Such a choice will then give a value d' which can be shown to be independent of the particular choice of x_1 and y_1, and the equation will reduce to one of the forms vii, viii, ix.

Thus the complete identification of any equation of the form (1) can be made from a consideration of the matrices **A** and **Q**, and will depend on the ranks of these matrices and the characteristic equation of **Q**. In any case the matrix **S** can be found which will reduce the equation (1) to one of the standard forms.

8.9 **Second degree Surfaces:** We now consider the second degree equation in three variables,

$$(1) \qquad ax^2 + by^2 + cz^2 + 2fyz + 2gxz + 2hxy + 2px + 2qy + 2rz + d = 0\,,$$

or in homogeneous coordinates,

$$(2) \qquad ax^2 + by^2 + cz^2 + 2fyz + 2gxz + 2hxy + 2pxw + 2qyw + 2rzw + dw^2 = 0\,.$$

Equation (2) may be written as

$$(3) \qquad (x\ y\ z\ w)\begin{pmatrix} a & h & g & p \\ h & b & f & q \\ g & f & c & r \\ p & q & r & d \end{pmatrix}\begin{pmatrix} x \\ y \\ z \\ w \end{pmatrix} = (0).$$

Writing

$$\mathbf{V} = \begin{pmatrix} x \\ y \\ z \\ w \end{pmatrix}, \quad \mathbf{V}_1 = \begin{pmatrix} x' \\ y' \\ z' \\ w' \end{pmatrix}, \quad \mathbf{S} = \begin{pmatrix} a_{11} & a_{12} & a_{13} & x_1 \\ a_{21} & a_{22} & a_{23} & y_1 \\ a_{31} & a_{32} & a_{33} & z_1 \\ 0 & 0 & 0 & 1 \end{pmatrix}, \quad \mathbf{A} = \begin{pmatrix} a & h & g & p \\ h & b & f & q \\ g & f & c & r \\ p & q & r & d \end{pmatrix},$$

we have $\mathbf{V} = \mathbf{S}\,\mathbf{V}_1$, so that $\mathbf{V}' = \mathbf{V}'_1\mathbf{S}'$, and writing (3) as $\mathbf{V}'\mathbf{A}\mathbf{V} = \mathbf{O}$, we get the transformed equation $\mathbf{V}_1'\mathbf{B}\mathbf{V}_1 = \mathbf{O}$ by writing

$$\mathbf{V}'\mathbf{A}\mathbf{V} = \mathbf{V}_1'\mathbf{S}'\mathbf{A}\mathbf{S}\mathbf{V}_1 = \mathbf{V}_1'\mathbf{B}\mathbf{V}_1\,,$$

where $\qquad \mathbf{B} = \mathbf{S}'\mathbf{A}\mathbf{S}\,.$

Here **S** has the form indicated in 8.6.

Again we write

$$\mathbf{Q} = \begin{pmatrix} a & h & g \\ h & b & f \\ g & f & c \end{pmatrix},$$

the matrix obtained from the quadratic terms of (1), and

$$\mathbf{T} = \begin{pmatrix} a_{11} & a_{12} & a_{13} \\ a_{21} & a_{22} & a_{23} \\ a_{31} & a_{32} & a_{33} \end{pmatrix},$$

so that **T** is unitary. Also, if the transformed equation of (1) is

$$(1') \qquad a'x'^2 + b'y'^2 + c'z'^2 + 2f'y'z' + 2g'x'z' + 2h'x'y' + 2p'x' + 2q'y' + 2r'z' + d' = 0 ,$$

we write

$$\mathbf{R} = \begin{pmatrix} a' & h' & g' \\ h' & b' & f' \\ g' & f' & c' \end{pmatrix}.$$

Then $\mathbf{R} = \mathbf{T}'\mathbf{Q}\mathbf{T} = \mathbf{T}^{-1}\mathbf{Q}\mathbf{T}$, **R** is independent of x_1, y_1, z_1, and the rank and characteristic equation of **R** are the same as those of **Q**, while the rank and determinant of **B** are the same as those of **A**. Choosing **T** so that **R** is in diagonal form, the equation (1) becomes

$$(1'') \qquad a'x'^2 + b'y'^2 + c'z'^2 + 2p'x' + 2q'y' + 2r'z' + d' = 0 .$$

We now enumerate the seventeen varieties of standard forms of equations of the second degree. They may be written and identified as follows:

i $\quad \dfrac{x^2}{a^2} + \dfrac{y^2}{b^2} + \dfrac{z^2}{c^2} - 1 = 0,$ the real ellipsoid (Fig. 18),

ii $\quad \dfrac{x^2}{a^2} + \dfrac{y^2}{b^2} + \dfrac{z^2}{c^2} + 1 = 0,$ the imaginary ellipsoid,

iii $\quad \dfrac{x^2}{a^2} + \dfrac{y^2}{b^2} + \dfrac{z^2}{c^2} = 0,$ the imaginary elliptic cone,

iv $\quad \dfrac{x^2}{a^2} + \dfrac{y^2}{b^2} - \dfrac{z^2}{c^2} + 1 = 0,$ the hyperboloid of two sheets (Fig. 19),

v $\quad \dfrac{x^2}{a^2} + \dfrac{y^2}{b^2} - \dfrac{z^2}{c^2} - 1 = 0,$ the hyperboloid of one sheet (Fig. 20),

vi $\quad \dfrac{x^2}{a^2} + \dfrac{y^2}{b^2} - \dfrac{z^2}{c^2} = 0,$ the real elliptic cone (Fig. 21),

vii $\quad \dfrac{x^2}{a^2} + \dfrac{y^2}{b^2} + 2rz = 0,$ the elliptic paraboloid (Fig. 22),

viii $\quad \dfrac{x^2}{a^2} - \dfrac{y^2}{b^2} + 2rz = 0,$ the hyperbolic paraboloid (Fig. 23),

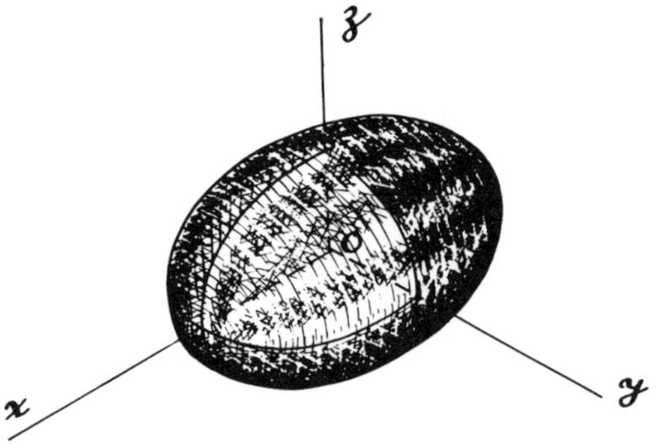

Fig 18

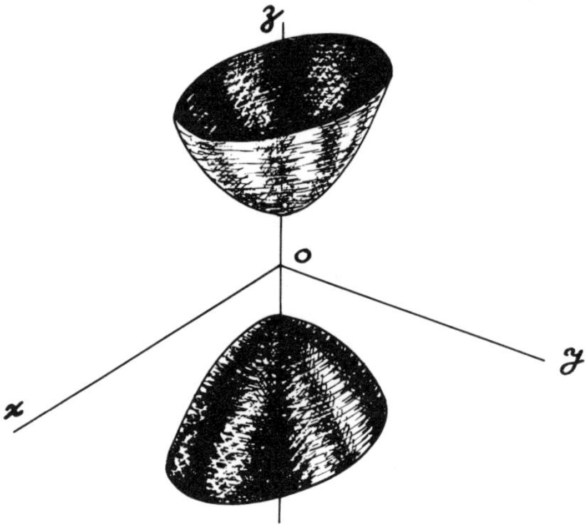

Fig. 19

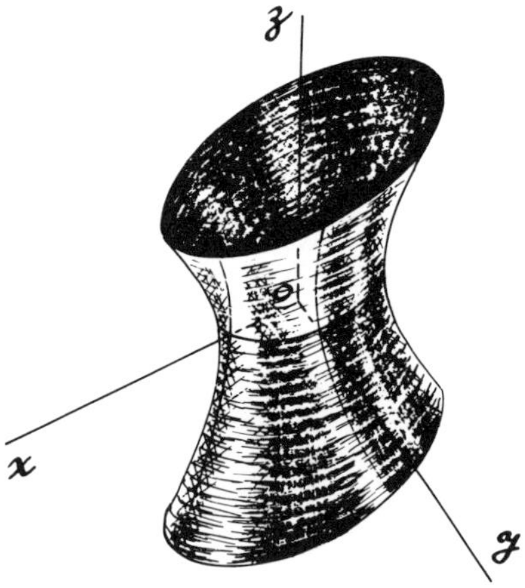

Fig. 20

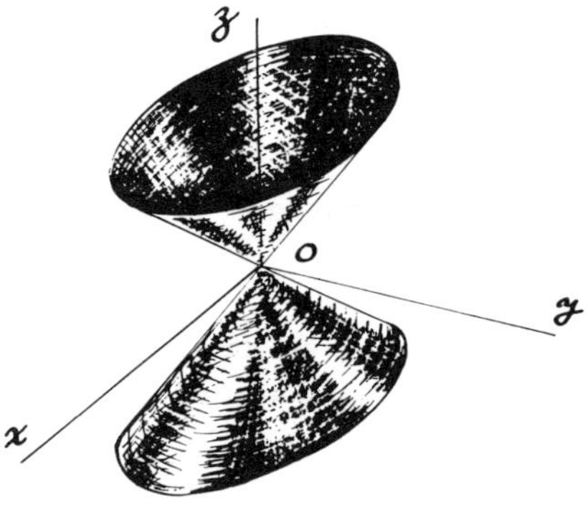

Fig. 21

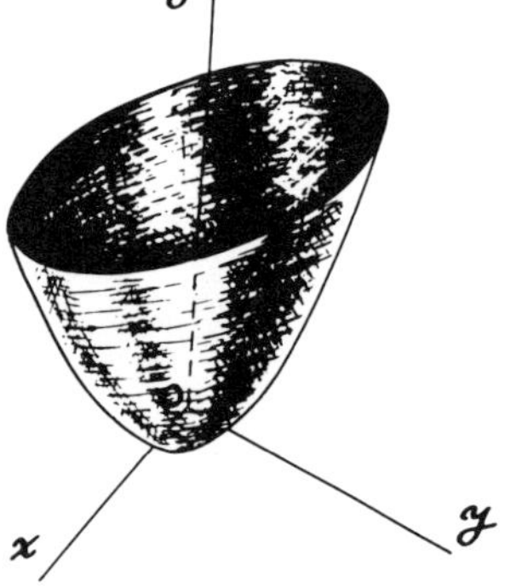

Fig. 22

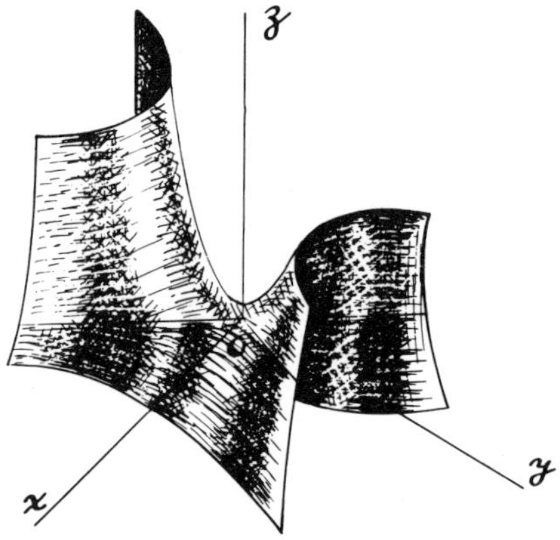

Fig. 23

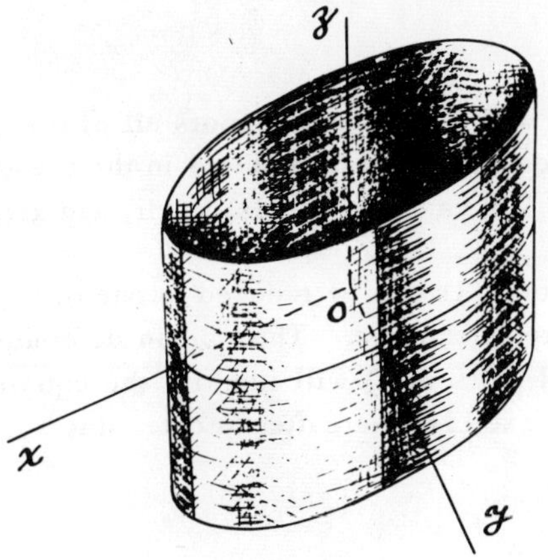

Fig 24

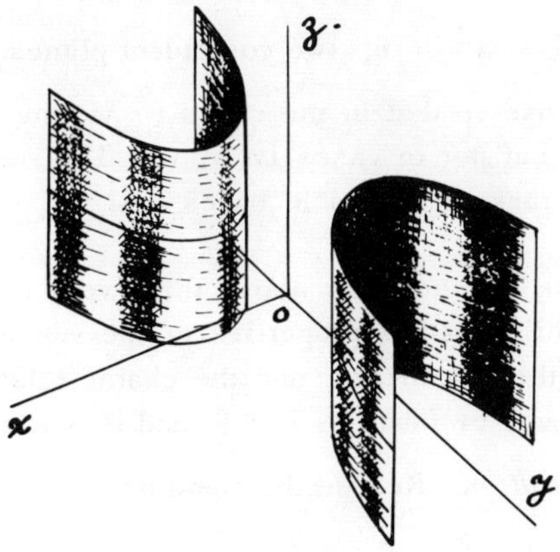

Fig. 25

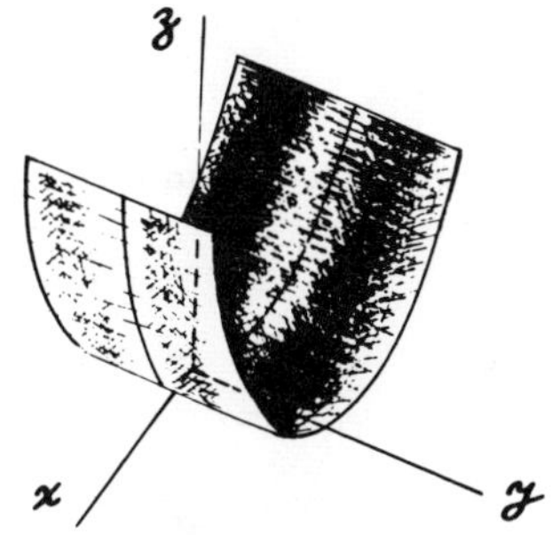

Fig. 26

ix $\quad \dfrac{x^2}{a^2} + \dfrac{y^2}{b^2} - 1 = 0,$ the real elliptic cylinder (Fig. 24),

x $\quad \dfrac{x^2}{a^2} + \dfrac{y^2}{b^2} + 1 = 0,$ the imaginary elliptic cylinder,

xi $\quad \dfrac{x^2}{a^2} + \dfrac{y^2}{b^2} = 0,$ imaginary intersecting planes,

xii $\quad \dfrac{x^2}{a^2} - \dfrac{y^2}{b^2} + 1 = 0,$ the hyperbolic cylinder (Fig. 25),

xiii $\quad \dfrac{x^2}{a^2} - \dfrac{y^2}{b^2} = 0,$ real intersecting planes,

xiv $\quad x^2 + 2\,q\,y = 0,$ the parabolic cylinder (Fig. 26),

xv $\quad x^2 - a^2 = 0,$ real parallel planes,

xvi $x^2 + a^2 = 0$, imaginary parallel planes,

xvii $x^2 = 0$, two coincident planes.

We observe that in the cases i - vi, the rank of **Q** is 3, with characteristic roots all of the same sign in i, ii, iii, but not in cases iv, v, vi. The rank of **Q** is 2 in cases vii - xiii, and is 1 in the remaining cases. Also the rank of **A** is 4 in cases i, ii, iv, v, vii, viii; it is 3 in cases iii, vi, ix, x, xii, and xiv, and 1 only in xvii.

It can be shown as above that every equation of the form (1) can be reduced to one of these standard forms, while various properties of the matrices **A** and **Q** are unchanged. Thus again determination of the ranks of these matrices, and the characteristic equation and roots of **Q** will identify the equation (1). The matrix **T** may be found as in 8.8, and it is a simple algebraic exercise thereafter to determine x_1, y_1, and z_1.

Illustration: Reduce the equation

$$x^2 + y^2 - 7z^2 - 2xy - 4xz - 4yz + 8y + 14z - 6 = 0$$

to standard form.

Let us first find the unitary transformation **T** which reduces the quadratic terms. The matrix **Q** is

$$\begin{pmatrix} 1 & -1 & -2 \\ -1 & 1 & -2 \\ -2 & -2 & -7 \end{pmatrix} .$$

The characteristic values of **Q** are found to be 1, 2, –8, and we find, as in illustration 5.13, that

$$\mathbf{T} = \begin{pmatrix} \frac{2}{3} & \frac{-\sqrt{2}}{2} & \frac{\sqrt{2}}{6} \\ \frac{2}{3} & \frac{\sqrt{2}}{2} & \frac{\sqrt{2}}{6} \\ -\frac{1}{3} & 0 & \frac{2\sqrt{2}}{3} \end{pmatrix} .$$

Then

$$\mathbf{S} = \begin{pmatrix} \frac{2}{3} & \frac{-\sqrt{2}}{2} & \frac{\sqrt{2}}{6} & x_1 \\ \frac{2}{3} & \frac{\sqrt{2}}{2} & \frac{\sqrt{2}}{6} & y_1 \\ \frac{-1}{3} & 0 & \frac{2\sqrt{2}}{3} & z_1 \\ 0 & 0 & 0 & 1 \end{pmatrix} .$$

We may now choose x_1, y_1, and z_1 so that the matrix $\mathbf{B} = \mathbf{S}'\mathbf{A}\,\mathbf{S}$ is in diagonal form, say

$$\begin{pmatrix} 1 & 0 & 0 & 0 \\ 0 & 2 & 0 & 0 \\ 0 & 0 & -8 & 0 \\ 0 & 0 & 0 & d \end{pmatrix} ,$$

and the reduced equation is

$$x'^2 + 2y'^2 - 8z'^2 + d = 0 ,$$

where d is the term obtained by the matrix multiplication above. This is not a convenient way generally to carry out the reduction. Simpler methods will be examined in the next chapter.

8.10 **Direction numbers and equations of straight lines and planes:** Let ξ and η be any two vectors with end points on a straight line d. We define $\delta = \xi - \eta$ to be a direction on d, and the components of $\xi - \eta$ are called the *direction numbers* of d. It is clear that a set of direction numbers is an ordered set, and we call these numbers respectively the x-direction, y-direction, and z-direction numbers.

It follows immediately that for a given straight line any two sets of direction numbers are proportional, and the equation of the straight line through $\mathbf{A} : (x_1, y_1, z_1)$ and $\mathbf{B} : (x_2, y_2, z_2)$ will be

$$(1) \qquad \begin{cases} x - x_1 = t(x_2 - x_1) \\ y - y_1 = t(y_2 - y_1) \\ z - z_1 = t(z_2 - z_1) \;. \end{cases}$$

We write (1) in the form

$$(2) \qquad \begin{cases} x = x_1 + \lambda\, t \\ y = y_1 + \mu\, t \\ z = z_1 + \nu\, t \;, \end{cases}$$

or in vector form

$$(3) \qquad \xi = \eta + t\,\delta, \text{ where } \eta = (x_1, y_1, z_1) \text{ and } \delta = (\lambda,\ \mu,\ \nu).$$

We call (2) or (3) a parametric equation of the line **A B**.

Now for a plane $\mathcal{P}$ let $\delta = (\lambda, \mu, \nu)$ be a direction perpendicular to $\mathcal{P}$ and $\mathbf{A} : (x_o,\ y_o,\ z_o)$ be a point of $\mathcal{P}$. Then an equation of $\mathcal{P}$ can be derived as follows:

Let $\mathbf{P}$: (x, y, z) be any point on $\mathcal{P}$. Then the vector $(x\text{-}x_o,\ y\text{-}y_o,\ z\text{-}z_o)$ is perpendicular to δ and we have

$$\lambda(x\text{-}x_o) + \mu\,(y\text{-}y_o) + \nu\,(z\text{-}z_o) = 0, \quad \text{or}$$

$$(\delta,\ \xi - \alpha) = 0, \text{ where } \xi = (x, y, z) \text{ and } \alpha = (x_o, y_o, z_o) \;.$$

8.11 **Intersection of a straight line and a quadric:** If we take homogeneous coordinates $(x, y, z, 1)$ for a vector $\xi = (x, y, z)$, we see by 8.10 that the homogeneous coordinates of a direction $\delta = (\lambda,\ \mu,\ \nu)$ will be $(\lambda,\ \mu,\ \nu,\ 0)$.

The equation of a quadric [see 8.9 (1)] can be written as

$$(1) \qquad (\mathbf{A}\,\xi,\ \xi) = 0 \;,$$

where $\xi = (x, y, z, 1)$, and $\mathbf{A}$ is the same matrix as in 8.9. Let

$$(2) \qquad \xi = \eta + t\,\delta \;,$$

be a straight line where $\eta = (x_1, y_1, z_1, 1)$ is fixed and δ is a fixed direction $(\lambda, \mu, \nu, 0)$, [see 8.10 (3)].

The points of intersection of (1) and (2) are found by the following second degree equation in t,

$$(3) \qquad (\mathbf{A}\delta,\ \delta)\, t^2 + 2t\,(\mathbf{A}\eta,\ \delta) + (\mathbf{A}\eta,\ \eta) = 0$$

Since $\delta = (\lambda, \mu, \nu, 0)$ we have $(\mathbf{A}\delta,\ \delta) = (\mathbf{Q}\delta,\ \delta)$, where $\mathbf{Q}$ is the matrix given in 8.9. Therefore (3) is written as

$$(4) \qquad (\mathbf{Q}\delta,\ \delta)\, t^2 + 2\,(\mathbf{A}\eta,\ \delta)\, t + (\mathbf{A}\eta,\ \eta) = 0 \;.$$

The following cases can be observed:

(a) If $(\mathbf{Q}\delta, \delta) \neq 0$, the line intersects the quadric in two points, real or complex conjugate, or coincident. This is determined through the discriminant of (4). The line segment connecting two points of a quadric is called a chord of the quadric.

(b) If $(\mathbf{Q}\delta, \delta) = 0$, $(\mathbf{A}\eta, \delta) \neq 0$, the line and quadric intersect in one point.

(c) If $(\mathbf{Q}\delta, \delta) = 0$, $(\mathbf{A}\eta, \delta) = 0$, but $(\mathbf{A}\eta, \eta) \neq 0$, the line and quadric do not intersect.

(d) If (4) is an identity, then the line lies on the quadric and this line is called a *ruling* of the quadric.

8.12 **Theorem:** (*Diametral plane*) The midpoints of all chords of a quadric parallel to a given direction δ lie in a plane called the *diametral plane* corresponding (*conjugate*) to that direction.

Proof: In 8.11 (4) let $(\mathbf{Q}\delta, \delta) \neq 0$. Then $\dfrac{t_1 + t_2}{2} = \dfrac{1}{(\mathbf{Q}\delta, \delta)}(\mathbf{A}\eta, \delta)$. Choosing η to end at the midpoint of the chord we get $t_1 = -t_2$ and therefore

$$(\mathbf{A}\eta, \delta) = 0 \ .$$

But we have $(\mathbf{A}\eta, \delta) = (\mathbf{A}\delta, \eta)$ because $\mathbf{A}$ is symmetric and the space is real. Thus when the end point of η varies over the midpoints of all chords with direction δ, we have $(\mathbf{A}\delta, \eta) = 0$. Considering

$$\delta = (\lambda, \mu, \nu, 0) \text{ and } \eta = (x, y, z, 1) \text{ and}$$

$$\mathbf{A} = \begin{pmatrix} a & h & g & p \\ h & b & f & q \\ g & f & c & r \\ p & q & r & d \end{pmatrix}, \text{ we get}$$

(1) $$(a\lambda + h\mu + g\nu)x + (h\lambda + b\mu + f\nu)y + (g\lambda + f\mu + c\nu)z + (p\lambda + q\mu + r\nu) = 0 \ .$$

This proves the theorem. It is convenient to write (1) in the form

(2) $$(\lambda\ \mu\ \nu\ 0) \begin{pmatrix} a & h & g & p \\ h & b & f & q \\ g & f & c & r \\ p & q & r & d \end{pmatrix} \begin{pmatrix} x \\ y \\ z \\ 1 \end{pmatrix} = (0) \ .$$

If the diametral plane is perpendicular to the corresponding direction δ, then δ is a characteristic direction of $\mathbf{Q}$. This is easily seen from

$$\begin{cases} a\lambda + h\mu + g\nu = k\lambda \\ h\lambda + b\mu + f\nu = k\mu \\ g\lambda + f\mu + c\nu = k\nu \ , \end{cases}$$

or

$$\mathbf{Q}\delta = k\delta \ .$$

In this case the diametral plane is called a *principal plane* of the quadric.

Illustration: Find the principal planes of

$$x^2 + y^2 - 7z^2 - 2xy - 4yz - 4xz + 8y + 14z - 6 = 0 \ .$$

The proper values of the matrix $\mathbf{Q}$ are 1, 2, –8, and the corresponding directions or proper vectors are (2,2,–1), (–1,1,0), and (1,1,4) [see 8.9, illustration].

For direction (2,2,–1), from equation (2) we get

$$(2\ 2\ {-1}\ 0)\begin{pmatrix} 1 & -1 & -2 & 0 \\ -1 & 1 & -2 & 4 \\ -2 & -2 & -7 & 7 \\ 0 & 4 & 7 & -6 \end{pmatrix}\begin{pmatrix} x \\ y \\ z \\ 1 \end{pmatrix} = (0)\ ,$$

or

$$2x + 2y - z + 1 = 0\ .$$

Similarly for (–1,1,0) we get

$$-2x + 2y + 4 = 0\ ,$$

and for (1,1,4) we have

$$-\ 8x - 8y - 32z + 32 = 0\ .$$

8.13 **A center of a quadric:** A point **P** : (x_o,y_o,z_o) is called a *center* of a quadric if for each point **M** of the quadric, there is another point **N** on the quadric such that $P = \frac{1}{2}(M + N)$.

Let $\eta = (x_o,y_o,z_o,1)$, and for **M** : (x,y,z) choose $\xi = (x,y,z,1)$. Then the line through **P** and **M** is

(1) $\qquad \xi = \eta + t\,\delta\ ,$

where $\delta = (\lambda,\ \mu,\ \nu,\ 0) = (x-x_o, y-y_o, z-z_o, 0)$. As in 8.11 the points of intersection of (1) and the quadric $(\mathbf{A}\,\xi,\ \xi) = 0$ are found by the equation

$$(\mathbf{Q}\delta,\ \delta)\ t^2 + 2(\mathbf{A}\eta,\ \delta)\ t + (\mathbf{A}\eta,\ \eta) = 0\ .$$

By 8.12 if **P** is the midpoint of the chord **MN**, then

$$(\mathbf{A}\eta,\ \delta) = 0, \quad \text{if} \quad (\mathbf{Q}\delta,\ \delta) \neq 0\ .$$

That is,

$$(ax_o + hy_o + gz_o + p)\,(x-x_o) + (hx_o + by_o + fz_o + q)\,(y-y_o) + (gx_o + fy_o + cz_o + r)\,(z-z_o) = 0\ .$$

This implies that the vector

$$\alpha = (ax_o + hy_o + gz_o + p,\ hx_o + by_o + fz_o + q,\ gx_o + fy_o + cz_o + r)$$

is perpendicular to $(x - x_o,\ y - y_o,\ z - z_o)$ for all (x,y,z). Therefore either the quadric is a double plane, or $\alpha = 0$, i.e., a center (x_o,y_o,z_o) is a solution of the system of equations

$$(2) \qquad \begin{cases} ax + hy + gz + p = 0 \\ hx + by + fz + q = 0 \\ gx + fy + cz + r = 0\ . \end{cases}$$

The system (2) may be written as

$$(3) \qquad (x\ y\ z\ 1)\begin{pmatrix} a & h & g \\ h & b & f \\ g & f & c \\ p & q & r \end{pmatrix} = (0\ 0\ 0)\ .$$

Illustration: Find the center of the quadric

$$x^2 + y^2 - 7z^2 - 2xy - 4xz - 4yz + 8y + 14z - 6 = 0 .$$

We solve the system

$$(x\ y\ z\ 1)\begin{pmatrix} 1 & -1 & -2 \\ -1 & 1 & -2 \\ -2 & -2 & -7 \\ 0 & 4 & 7 \end{pmatrix} = (0\ 0\ 0),$$

or

$$\begin{cases} x - y - 2z = 0 \\ -x + y - 2z + 4 = 0 \\ -2x - 2y - 7z + 7 = 0 \end{cases}$$

which gives (1,–1,1).

8.14 **Tangent plane to a quadric:** In 8.11 (4), for a double point we have

(1) $$(\mathbf{A}\eta,\ \delta)^2 - (\mathbf{Q}\delta,\ \delta)\,(\mathbf{A}\eta,\ \eta) = 0 .$$

If η ends at the point of tangency, we have $(\mathbf{A}\eta,\ \eta) = 0$ and δ will be the direction of a line tangent to the quadric. Therefore from (1) we have

(2) $$(\mathbf{A}\eta,\ \delta) = 0 .$$

Let $\eta = (x_o,\ y_o,\ z_o,\ 1)$ and $\delta = (\lambda,\ \mu,\ \nu,\ 0)$. Then (2) implies

$$(ax_o + hy_o + gz_o + p)\,\lambda + (hx_o + by_o + fz_o + q)\,\mu + (gx_o + fy_o + cz_o + r)\,\nu = 0 .$$

This means that the vector

$$\alpha = (ax_o + hy_o + gz_o + p,\ hx_o + by_o + fz_o + q,\ gx_o + fy_o + cz_o + r)$$

is perpendicular to δ and therefore to the locus of δ, i.e., the tangent plane of the quadric, and α is normal to the quadric. Thus the equation of the tangent plane will be

(3) $$(ax_o + hy_o + gz_o + p)\,(x-x_o) + (hx_o + by_o + fz_o + q)\,(y-y_o) + (gx_o + fy_o + cz_o + r)\,(z-z_o) = 0 .$$

Since (x_o,y_o,z_o) is a point of the quadric

$$ax_o^2 + by_o^2 + cz_o^2 + 2hx_oy_o + 2gx_oz_o + 2fy_oz_o + 2px_o + 2qy_o + 2rz_o + d = 0 ,$$

and we have

$$-ax_o^2 - by_o^2 - cz_o^2 - 2fy_oz_o - 2gx_oz_o - 2hx_oy_o - px_o - qy_o - rz_o = px_o + qy_o + rz_o + d .$$

Hence equation (3) may be written in matrix form as

$$(x_o\ y_o\ z_o\ 1)\begin{pmatrix} a & h & g & p \\ h & b & f & q \\ g & f & c & r \\ p & q & r & d \end{pmatrix}\begin{pmatrix} x \\ y \\ z \\ 1 \end{pmatrix} = (0) .$$

Illustration: Find equations of tangent plane and normal line to the quadric

$$2x^2 + 3y^2 - 2z^2 + 4xy + 6yz - 8xz + 12x - 10y + 8z - 3 = 0$$

at the point (1,–1,2).

The tangent plane is

$$(1\ -1\ 2\ 1)\begin{pmatrix} 2 & 2 & -4 & 6 \\ 2 & 3 & 3 & -5 \\ -4 & 3 & -2 & 4 \\ 6 & -5 & 4 & -3 \end{pmatrix}\begin{pmatrix} x \\ y \\ z \\ 1 \end{pmatrix} = (0) ,$$

or

$$-2x - 7z + 16 = 0 .$$

The normal line is

$$\begin{cases} x = 1 - 2t \\ y = -1 \\ z = 2 - 7t . \end{cases}$$

8.15 **Ruled surfaces:** A surface S is called a *ruled surface* if through every point of S there is a straight line which lies entirely in S. For a quadric $(A\xi, \xi) = 0$ as in 8.11, let $\xi = \eta + t\,\delta$ be a line passing through the end point of η, a point of the quadric. Then by 8.11 (d) the line will lie on $(\mathbf{A}\xi, \xi) = 0$ if

$$(1)\qquad \begin{cases} (\mathbf{Q}\delta, \delta) = 0 \\ (\mathbf{A}\eta, \delta) = 0 \\ (\mathbf{A}\eta, \eta) = 0 . \end{cases}$$

The third equality is already satisfied since the end point of η is on the quadric. Therefore δ is determined through the following equations:

$$(2)\qquad \begin{cases} (\mathbf{Q}\delta, \delta) = 0 \\ (\mathbf{A}\eta, \delta) = 0 . \end{cases}$$

By 8.14 (2) the equation of the tangent plane at the end point of η is $(\mathbf{A}\eta, \delta) = 0$. This proves that if a line lies in a quadric it lies in a tangent plane of the quadric.

Now let $\delta = (\lambda, \mu, \nu, 0)$ and $\eta = (x_o, y_o, z_o, 1)$.

Then (2) will be

$$(3)\qquad \begin{cases} (a\lambda + h\mu + g\nu)\,\lambda + (h\lambda + b\mu + f\nu)\,\mu + (g\lambda + f\mu + c\nu)\,\nu = 0 \\ (ax_o + hy_o + gz_o + p)\,\lambda + (hx_o + by_o + fz_o + q)\,\mu + (gx_o + fy_o + cz_o + r)\,\nu = 0 . \end{cases}$$

But λ, μ, ν being direction numbers, we can suppose that they are homogeneous coordinates of a point in a plane i.e., we let $x = \frac{\lambda}{\nu}$ and $y = \frac{\mu}{\nu}$ if $\nu \neq 0$; otherwise (λ, μ, ν) is a point at infinity. Therefore (3) is equivalent to finding points of intersection of a conic section and a straight line, and we will have the following cases:

(a) If we get two real solutions for (3), then through (x_o, y_o, z_o) of the quadric pass two real rulings.
(b) If (3) has a double solution, then through (x_o, y_o, z_o) passes a double ruling.

(c) If (3) has complex solutions, then the ruling is complex.

(d) For the trivial case that the quadric degenerates into planes we have infinitely many rulings.

It is customary to call a surface S a *ruled surface* if S has *real rulings*.

To consider the problem of rulings of a quadric with given direction we observe from (1) that this direction must be such that $(\mathbf{Q}\delta, \delta) = 0$, and the equations

$$(4) \qquad \begin{aligned} (\mathbf{A}\eta, \delta) &= 0 \\ (\mathbf{A}\eta, \eta) &= 0 \end{aligned}$$

will give all points through which the rulings pass. The cases of real and complex solutions should be considered as above.

Illustration 1: Find the rulings of the quadric

$$xy + 2xy + 3yz + y + z + 2 = 0$$

through the point $(1,-1,1)$ of the quadric.

Multiplying the equation by 2 for convenience, we get

$$2xy + 4xz + 6yz + 2y + 2z + 4 = 0 ,$$

and equations (2) are then

$$\lambda\,\mu + 2\,\lambda\,\nu + 3\,\mu\,\nu = 0$$

and

$$(1\ -1\ 1\ 1)\begin{pmatrix} 0 & 1 & 2 \\ 1 & 0 & 3 \\ 2 & 3 & 0 \\ 0 & 1 & 1 \end{pmatrix}\begin{pmatrix} \lambda \\ \mu \\ \nu \end{pmatrix} = (0)$$

or $\qquad \lambda + 5\,\mu = 0.$

This gives either $(0,0,1)$ or $(35,-7,5)$. Thus the rulings are

$$\begin{cases} x = 1 \\ y = -1 \\ z = 1 + t\ , \end{cases} \qquad \begin{cases} x = 1 + 35t \\ y = -1 - 7t \\ z = 1 + 5t\ . \end{cases}$$

Illustration 2: Find the rulings of the quadric

$$xy + 2xz + 3yz + y + z + 2 = 0$$

having the direction $(1,-1,-1)$.

We observe that for $\delta = (1,-1,-1)$, $(\mathbf{Q}\delta, \delta) = 0$, so the first equation of (1) is satisfied. We must then find an $\eta = (x,y,z)$ satisfying equations (4). This yields

$$(x\ y\ z\ 1)\begin{pmatrix} 0 & 1 & 2 \\ 1 & 0 & 3 \\ 2 & 3 & 0 \\ 0 & 1 & 1 \end{pmatrix}\begin{pmatrix} 1 \\ -1 \\ -1 \end{pmatrix} = (0) ,$$

or

$$3x + 2y + z + 2 = 0 ,$$

and

$$xy + 2xz + 3yz + y + z + 2 = 0 .$$

Since we need only one point to determine a ruling, let us find where the ruling cuts some plane which is not parallel to it, say $z = 0$. We must then solve the equations

$$\begin{cases} 3x + 2y + 2 = 0 \\ xy + y + 2 = 0 . \end{cases}$$

We get two points,

$$\left(\frac{1}{3}, -\frac{3}{2}, 0\right) \quad \text{and} \quad (2,-2,0) .$$

Thus there are two rulings with the given direction, namely

$$\begin{cases} x = \frac{1}{3} + t \\ y = -\frac{3}{2} - t \\ z = -t \end{cases}, \quad \text{and} \quad \begin{cases} x = 2 + t \\ y = -2 - t \\ z = -t \end{cases} .$$

8.16 **Theorem:** (*Pole and polar*) Let any line through $P : (x_o,y_o,z_o)$ intersect a quadric in two points $B : (x_1,y_1,z_1)$ and $C : (x_2,y_2,z_2)$. A point D is called the harmonic conjugate of P with respect to B and C if

$$\frac{PB}{PC} = -\frac{DB}{DC} ,$$

where, for example PB is the directed distance from P to B. The locus of D is a plane called the *polar* of P with respect to the quadric.

Proof: Let η end at P. Then for any direction δ, the equation of a line through P is:

(1) $$\xi = \eta + t\,\delta .$$

The points of intersection of (1) and the quadric $(\mathbf{A}\xi, \xi) = 0$ are obtained by

$$(\mathbf{Q}\delta, \delta)t^2 + 2(\mathbf{A}\eta, \delta)t + (\mathbf{A}\eta, \eta) = 0$$

[see 8.11 (4)]. Let $(\mathbf{Q}\delta, \delta) \neq 0$. Clearly $t = 0$ corresponds to P. Let t_1 and t_2 correspond to B and C, and t correspond to D. Then for D to be the harmonic conjugate of P with respect to B and C, we have

$$\frac{0 - t_1}{0 - t_2} = -\frac{t - t_1}{t - t_2} .$$

This gives

$$t = \frac{2t_1t_2}{t_1 + t_2} = -\frac{(\mathbf{A}\eta, \eta)}{(\mathbf{A}\eta, \delta)} \quad \text{if } (\mathbf{A}\eta, \delta) \neq 0 .$$

Substituting this in (1) we get

$$(\mathbf{A}\eta, \delta)\,\xi = (\mathbf{A}\eta, \delta)\,\eta - (\mathbf{A}\eta, \eta)\,\delta .$$

Taking the inner product of both sides by $\mathbf{A}\,\eta$ we get

$$(\mathbf{A}\eta, \delta)(\mathbf{A}\eta, \xi) = (\mathbf{A}\eta, \delta)(\mathbf{A}\eta, \eta) - (\mathbf{A}\eta, \eta)(\mathbf{A}\eta, \delta) = 0 .$$

Since $(\mathbf{A}\eta, \delta) \neq 0$ we have

$$(\mathbf{A}\eta, \xi) = 0$$

which is the equation of a plane (the polar of P with respect to the quadric).

If $(\mathbf{A}\eta, \delta) = 0$, the point P is a center of the quadric, therefore the midpoint of the segment BC. In this case the harmonic conjugate is not defined.

If $\mathcal{p}$ is the polar of a point P with respect to a quadric, then P is called the *pole* of $\mathcal{p}$ with respect to the quadric.

EXERCISES 8

1. Reduce to cononical form

   (i) $x^2 + 6xy - 2y^2 - 3yz + z^2$,
   (ii) $-2x^2 - 11y^2 - 5z^2 + 4xy + 16yz + 20xz$,
   (iii) $3x^2 - y^2 - 3z^2 + 3t^2 - 4xy - 10yt$.

2. Find the matrix of the change of base in each part of exercise 1.
3. Reduce the following conic sections to one of the nine forms of 8.8.

   (i) $52x^2 - 72xy + 73y^2 + 8x - 294y - 1167 = 0$,
   (ii) $66x^2 - 24xy + 59y^2 + 156x - 142y + 149 = 0$,
   (iii) $16x^2 + 24xy + 9y^2 + 150x - 200y - 1000 = 0$,
   (iv) $3x^2 + 8xy - 3y^2 + 54x + 22y - 77 = 0$,
   (v) $144x^2 + 120xy + 25y^2 - 260x + 624y + 676 = 0$,
   (vi) $108x^2 + 300xy - 17y^2 + 1116x + 198y + 855 = 0$,
   (vii) $64x^2 - 96xy + 36y^2 + 480x - 360y + 675 = 0$.

4. By the method used in 8.13 find the center (if any) of the conic sections in 3.
5. By a method similar to the one in 8.14 find the equation of the tangent line at a point (x_o, y_o) to the conic sections in 3.
6. Find the points of intersection of

$$\begin{cases} x = 1 + 2t \\ y = -2 + t \\ z = 1 - 3t \end{cases},$$

   and

$$x^2 + 6y^2 + 4yz + 2xy + 8y + z - 2 = 0 .$$

7. Discuss the following surfaces. Draw a diagram.

   (i) $y^2 - 9z^2 = 81$, (ii) $4x^2 + 4y^2 + z^2 = 16$,
   (iii) $x^2 + y^2 + z^2 + 2xy + 2xz + 2yz = 1$, (iv) $4x = 4y^2 - z^2$.

8. Find the diametral plane of the given surface corresponding to the given direction.

   (i) $3x^2 - y^2 + 4z^2 + 5 = 0$; $(2,4,-3)$,
   (ii) $xy + xy + yz = 1$; $(1,1,1)$,
   (iii) $z^2 + 2yx + 6z - 12z - 3 = 0$; $(3,-1,2)$.

9. Find the centers of the following quadrics

   (i) $xy + yz + x + 2z = 0$,
   (ii) $x^2 + 4y^2 + 4xy + 6xz + 12yz - 8x - 16y - 6z = 0$.

10. Write the equations of the tangent plane and normal line of each given quadric at the given point.

(i) $5x^2 + y^2 + 2z^2 = 49$; $(1,-6,2)$,
(ii) $2x^2 + z^2 + 6xy - 2xz + 10y + 12z - 25 = 0$; $(1,2,-1)$,
(iii) $x^2 + 6yz + 4x - 2y + 14z = 0$; $(0,0,0)$.

11. Determine which ones of the seventeen quadrics of 8.9 are ruled surfaces with real, imaginary, or double rulings. In case of real rulings, find the equations of the rulings at any point (x_o,y_o,z_o) of the surface.

12. Write the equations of the rulings of

(i) $x^2 + y^2 - z^2 = 1$ at $(1,1,1)$,
(ii) $\frac{x^2}{9} + \frac{y^2}{4} - z^2 = 1$ at $(12,-14,8)$,
(iii) $xy + 2xz + 3yz + y + z + 2 = 0$ at $(1,-1,1)$.

13. Find the equations of the rulings of the following surfaces having the given directions:

(i) $x^2 + y^2 - 1 = 0$, direction $(0,0,1)$,
(ii) $\frac{x^2}{16} + \frac{y^2}{4} - z^2 = 1$, direction $(16,-6,5)$,
(iii) $\frac{x^2}{25} - y^2 + 2z = 0$, direction $(5,1,-6)$,
(iv) $x^2 + y^2 - z^2 - 1 = 0$, direction $(3,4,5)$.

14. Show that through any point of the surface

$$13x^2 + 10y^2 + 5z^2 - 4xy - 6xz - 12\,yz - 56 = 0$$

there is only one real ruling. Find its direction.

15. By the method of 8.16 find the polar of $P : (x_o,y_o)$ with respect to a given conic section.

16. Find the polar of the given point with respect to the given quadric:

(i) $3x^2 + y^2 - z^2 + 2xy - 4yz + 6x - 10y + 7 = 0$; $(1,1,1)$,
(ii) $x^2 - 5z^2 + 4xy - 8xz + 10y - 2 = 0$; $(0,-1,2)$,
(iii) $y^2 - z^2 + 6xy - 12xz + 8yz - 16x = 0$; $(1,-1,3)$.

ADDITIONAL PROBLEMS 8

1. Show that the line $x - y = 1$ is a part of the conic

$$x^2 - xy - y - 1 = 0 .$$

2. Find the points of intersection of line and conic:

(i) $\begin{cases} x = 1 + t \\ y = 2 - t \end{cases}$; $x^2 - 8xy + y^2 - 2x - 2y + 1 = 0$,
(ii) $y = x$; $x^2 - 2xy + y^2 - 2x - 2y + 1 = 0$,
(iii) $y = 3x - 1$, $(x\ y\ 1)\begin{pmatrix} 1 & 7 & 0 \\ 7 & 2 & 2 \\ 0 & 2 & 3 \end{pmatrix}\begin{pmatrix} x \\ y \\ 1 \end{pmatrix} = (0)$,
(iv) $2x + y = 7$, $2x^2 - 2xy - y^2 - 8x + 2 = 0$.

3. Find the points of intersection of line and quadric:

(i) $\begin{cases} x = 1 + t \\ y = 2 - t \\ z = -2t \, , \end{cases}$ $x^2 + 2y^2 - 3z^2 + 2xy - 2yz + 4xz - 6x - 10z + 6 = 0$,

(ii) $x - 3 = \dfrac{y+1}{-1} = \dfrac{z+4}{-3}$, $xy + yz + xz - 2x - 6y + 3 = 0$,

(iii) $\begin{cases} x = 2 \\ y = 2 + 2t \\ z = 1 + t \, , \end{cases}$ $xy - 4z = 0$.

4. Find the centers of the following conics:

(i) $52x^2 - 72xy + 73y^2 + 8x - 294y - 1167 = 0$,
(ii) $16x^2 + 24xy + 9y^2 + 150x - 200y - 1000 = 0$,
(iii) $64x^2 - 96xy + 36y^2 + 480x - 360y + 675 = 0$,
(iv) $17x^2 - 6xy + 9y^2 - 126x + 90y + 81 = 0$,
(v) $24xy - 7y^2 - 120y - 144 = 0$.

5. Find the centers of the following quadrics:

(i) $3x^2 + 2y^2 + 2z^2 - 4xy - 6xz - 8yz - 10x - 12y + 14z - 16 = 0$,
(ii) $13x^2 + 10y^2 + 5z^2 - 4xy - 6xz - 12yz - 56 = 0$,
(iii) $12x^2 - z^2 + 6xy - 8yz + 2x - 6y - 12z + 14 = 0$.

6. Find the diameter of the conic corresponding to the given direction:

(i) $3x^2 - 8xy + y^2 - 2x - 4y = 0$, direction (1,2) ,
(ii) $x^2 + 4y^2 - 64 = 0$, direction (–3,2) ,
(iii) $x^2 - 2xy - 7y^2 - 8x + 1 = 0$, direction (0,1) .

7. Find the diametral plane of the quadric corresponding to the given direction:

(i) $x^2 + 2y^2 - z^2 - 6xy - 4xz + 4yz - 8x + 6y + 6z - 12 = 0$, (1,1,2) ,
(ii) $x^2 + 4y^2 - 3z^2 + 10xy - 6y - 2z + 2 = 0$, (1,–1,1) ,
(iii) $6x^2 - z^2 + 10xy - 6xz + 6y + 8z - 5 = 0$, (3,1,–2) .

8. Find the axes of symmetry of the conics:

(i) $x^2 - 2xy + y^2 - 2x + 1 = 0$,
(ii) $x^2 - 4xy + 4y^2 - 10x + 8y = 0$,
(iii) $24xy - 7y^2 - 12y - 144 = 0$.

9. Find the vertex of the conics (i) and (ii) of problem 8, without changing coordinates.
10. Find the principal planes of the quadrics

(i) $2x^2 + 18y^2 + 2z^2 + 12xy + 2xz + 6yz - 2x - 8y + 7 = 0$,
(ii) $4x^2 + 9y^2 + 16z^2 - 12xy + 16xz - 24yz + 4x - 4y + 24z + 20 = 0$,
(iii) $3x^2 + 11y^2 - 7z^2 + 6xy + 2xz + 6yz + 6x - 2z + 1 = 0$.

11. Find the equations of the tangent and normal to the conic at the given point:

(i) $x^2 - 16xy + 4y^2 - 9x - 1 = 0$; (1,–1/2) ,
(ii) $\dfrac{x^2}{16} + \dfrac{y^2}{25} = 1$; (16/5,3) ,
(iii) $3xy - 7y^2 - 27y + 28 = 0$; (9,2) .

12. Find the equations of the tangent plane and normal line to the quadric at the given point:

(i) $x^2 + y^2 + z^2 - 4xy + 8yz - 6x - 1 = 0$; $(1,1,1)$,

(ii) $13x^2 + 10y^2 + 5z^2 - 4xy - 6xz - 12yz - 56 = 0$; $\left(\frac{2}{\sqrt{3}}, \frac{2}{\sqrt{3}}, \frac{-2}{\sqrt{3}}\right)$,

(iii) $xy + yz - 2xz - 6x - 8y - 7 = 0$; $(1,-1,-2)$.

13.* Write the equation of a plane tangent to the given quadric and perpendicular to the given direction:

(i) $x^2 + 2y^2 - 3z^2 + 4xy + 6yz - 2x + 4y - 3 = 0$; $(1,-1,-2)$,
(ii) $3y^2 + z^2 - 4xy - 4yz + 2x + 2y + 2z = 0$; $(1,0,1)$.

14.* Find equations of the two planes containing the line

$$\begin{cases} x = 1 + t \\ y = 1 - t \\ z = 1 \end{cases}$$

and tangent to $\frac{x^2}{9} + \frac{y^2}{4} + z^2 = 1$.

15. Find the rulings of the given quadric through the given point:

(i) $x^2 + y^2 - z^2 = 0$; $(3,4,5)$,
(ii) $x^2 - 2y^2 - 2xy - 4yz - 2z - 1 = 0$; $(-1,1,0)$.

16. Find the pole of the given plane with respect to the given quadric:

(i) $x^2 + 2y^2 + 3z^2 - 5 = 0$; $x + y + z = 10$,
(ii) $x^2 - 3z^2 + 2xy + 6yz - 6x + 10y - 4 = 0$; $3x + y - 2z = 1$.

17. Show that if a point moves on the line

$$\begin{cases} x = -1 - 2t \\ y = 2 + t \\ z = 3 + 2t \end{cases} ,$$

the polar of it with respect to the quadric

$$x^2 + 3z^2 - 2xy - 4yz - 4y - 49 = 0$$

revolves about a straight line.

18. State and prove the general theorem of which problem 17 is an example.

9. APPLICATIONS AND PROBLEM SOLVING TECHNIQUES

9.1 **A general projection:** Show that in a real three-dimensional space the projection **A** on the plane

(1) $ax + by + cz = 0$

is a linear transformation, and find the matrix of this transformation.

Solution: Let $V = (x_o, y_o, z_o)$. The projection of V on the plane is the vector $\mathbf{A}\,V = W$, connecting the origin to the point of intersection of the line

$$(2) \qquad \begin{cases} x = x_o + at \\ y = y_o + bt \\ z = z_o + ct \end{cases}$$

and the plane. Substituting (2) in (1) we get

$$t = -\frac{ax_o + by_o + cz_o}{a^2 + b^2 + c^2}, \quad \text{and therefore}$$

$$W = \left(x_o - \frac{a^2x_o + aby_o + acz_o}{a^2 + b^2 + c^2},\ y_o - \frac{abx_o + b^2y_o + bcz_o}{a^2 + b^2 + c^2},\ z_o - \frac{acx_o + bcy_o + c^2z_o}{a^2 + b^2 + c^2}\right).$$

It is clear that **A** is linear and the matrix of **A** is

$$\begin{pmatrix} 1 - \dfrac{a^2}{a^2 + b^2 + c^2} & \dfrac{-ab}{a^2 + b^2 + c^2} & \dfrac{-ac}{a^2 + b^2 + c^2} \\ \dfrac{-ab}{a^2 + b^2 + c^2} & 1 - \dfrac{b^2}{a^2 + b^2 + c^2} & \dfrac{-bc}{a^2 + b^2 + c^2} \\ \dfrac{-ac}{a^2 + b^2 + c^2} & \dfrac{-bc}{a^2 + b^2 + c^2} & 1 - \dfrac{c^2}{a^2 + b^2 + c^2} \end{pmatrix}.$$

9.2 **Intersection of planes:** Let

$$a_{11}x + a_{12}y + a_{13}z = b_1,$$

$$a_{21}x + a_{22}y + a_{23}z = b_2,$$

$$a_{31}x + a_{32}y + a_{33}z = b_3$$

be equations of three planes, so that at least one coefficient in each equation is different from zero. Give the geometric properties of the planes corresponding to all possible ranks of the matrices

$$\mathbf{A} = \begin{pmatrix} a_{11} & a_{12} & a_{13} \\ a_{21} & a_{22} & a_{23} \\ a_{31} & a_{32} & a_{33} \end{pmatrix}, \text{ and } \mathbf{B} = \begin{pmatrix} a_{11} & a_{12} & a_{13} & b_1 \\ a_{21} & a_{22} & a_{23} & b_2 \\ a_{31} & a_{32} & a_{33} & b_3 \end{pmatrix}.$$

Solution: Let $r(\mathbf{A})$ denote the rank of **A**.

I $r(\mathbf{A}) = 1$, $r(\mathbf{B}) = 1$. Each row of **B** can be written as some c times its first row; the three planes coincide.

II $r(\mathbf{A}) = 1$, $r(\mathbf{B}) = 2$. Either the three planes are parallel or two coincide and the third is parallel to them.

III $r(\mathbf{A}) = 2$, $r(\mathbf{B}) = 2$. The three planes have a line in common, or two coincide and the third intersects it in a line.

IV $r(\mathbf{A}) = 2$, $r(\mathbf{B}) = 3$. The line of intersection of two of the planes is parallel to the third (two of the planes may be parallel, the third intersects both).

V $r(\mathbf{A}) = 3$, $r(\mathbf{B}) = 3$. The three planes meet in one and only one point.

9.3 **Sphere:** Write the equation of the sphere

(1) $$(x-a)^2 + (y-b)^2 + (z-c)^2 = r^2$$

in matrix form and then inner product form.

Solution: Let $\xi = (x\ y\ z\ 1)$, and

$$\mathbf{A} = \begin{pmatrix} 1 & 0 & 0 & -a \\ 0 & 1 & 0 & -b \\ 0 & 0 & 1 & -c \\ -a & -b & -c & a^2+b^2+c^2-r^2 \end{pmatrix} .$$

Then we have (1) as follows:

$$(x\ y\ z\ 1) \begin{pmatrix} 1 & 0 & 0 & -a \\ 0 & 1 & 0 & -b \\ 0 & 0 & 1 & -c \\ -a & -b & -c & a^2+b^2+c^2-r^2 \end{pmatrix} \begin{pmatrix} x \\ y \\ z \\ 1 \end{pmatrix} = (0) .$$

Also (1) has the form

$$(\mathbf{A}\xi, \xi) = 0 .$$

9.4 **A property of the sphere:** If ϵ_1, ϵ_2, and ϵ_3 are three orthonormal vectors, then there are scalars h_1, h_2, k_1, k_2, l_1, and l_2 such that $h_1\epsilon_1$, $h_2\epsilon_1$, $k_1\epsilon_2$, $k_2\epsilon_2$, $l_1\epsilon_3$, and $l_2\epsilon_3$ end on the sphere in 9.3. Show that

$$h_1^2 + h_2^2 + k_1^2 + k_2^2 + l_1^2 + l_2^2 = 6r^2 - 2(a^2 + b^2 + c^2) .$$

Solution: The equation of the line containing ϵ, a unit vector, is

(1) $$\xi = \eta + t\,\epsilon ,$$

where $\eta = (0,0,0,1)$, and $\epsilon = (\lambda,\mu,\nu,0)$. By 8.11 (4) we have

$$t^2 - 2(a\lambda + b\mu + c\nu)t + a^2 + b^2 + c^2 - r^2 = 0$$

for the intersection of (1) and the sphere.

Using $\epsilon_1 = (\lambda_1, \mu_1, \nu_1, 0)$ and $t = h$ we get

(2) $$\begin{cases} h_1 + h_2 = 2(a\lambda_1 + b\mu_1 + c\nu_1) \\ h_1 h_2 = a^2 + b^2 + c^2 - r^2 \end{cases} .$$

Similarly for $\epsilon_2 = (\lambda_2, \mu_2, \nu_2, 0)$ and $t = k$ we get

(3) $$\begin{cases} k_1 + k_2 = 2(a\lambda_2 + b\mu_2 + c\nu_2) \\ k_1 k_2 = a^2 + b^2 + c^2 - r^2 \end{cases} .$$

Also for $\epsilon_3 = (\lambda_3, \mu_3, \nu_3, 0)$ and $t = l$ we get

$$(4) \qquad \begin{cases} l_1 + l_2 = 2(a\lambda_3 + b\mu_3 + c\nu_3) \\ l_1 l_2 = a^2 + b^2 + c^2 - r^2 \end{cases} .$$

Combining (2), (3), and (4) and noting that

$$(a\lambda_1 + b\mu_1 + c\nu_1)^2 + (a\lambda_2 + b\mu_2 + c\nu_2)^2 + (a\lambda_3 + b\mu_3 + c\nu_3)^2 = a^2 + b^2 + c^2 ,$$

we get the proof.

9.5 **Radical axis:** Prove that the radical axis of two circles

$$(x - a_1)^2 + (y - b_1)^2 - r_1{}^2 \quad \text{and} \quad (x - a_2)^2 + (y - b_2)^2 - r_2{}^2 = 0$$

is the line

$$2(a_2 - a_1)x + 2(b_2 - b_1)y + a_1{}^2 - a_2{}^2 + b_1{}^2 - b_2{}^2 + r_2{}^2 - r_1{}^2 = 0 .$$

Solution: Let η end at $\mathbf{P}$, a point whose powers with respect to both circles are the same. Let $(\mathbf{A}_1 \xi, \xi) = 0$ and $(\mathbf{A}_2 \xi, \xi) = 0$ be respectively the equations of the circles. By 8.11 (4) the line $\xi = \eta + t\,\delta$ intersects the circle at the points for which $t = h_1$ and $t = h_2$, the roots of

$$t^2 + 2\,(\mathbf{A}_1 \eta, \delta)\,t + (\mathbf{A}_1 \eta, \eta) = 0 ,$$

where $|\delta| = 1$. Thus $h_1 h_2 = (\mathbf{A}_1 \eta, \eta)$. Similarly for the second circle if the roots are k_1 and k_2 we have $k_1 k_2 = (\mathbf{A}_2 \eta, \eta)$. Therefore if

$$h_1 h_2 = k_1 k_2 ,$$

i.e., the powers are equal,

$$([\mathbf{A}_1 - \mathbf{A}_2]\, \eta, \eta) = 0$$

or

$$(x\ y\ 1) \left[\begin{pmatrix} 1 & 0 & -a_1 \\ 0 & 1 & -b_1 \\ -a_1 & -b_1 & a_1^2 + b_1^2 - r_1^2 \end{pmatrix} - \begin{pmatrix} 1 & 0 & -a_2 \\ 0 & 1 & -b_2 \\ -a_2 & -b_2 & a_2^2 + b_2^2 - r_2^2 \end{pmatrix} \right] \begin{pmatrix} x \\ y \\ 1 \end{pmatrix} = (0) .$$

This proves the statement.

Slight generalization: Let

$$ax^2 + 2hxy + by^2 + 2p_1 x + 2q_1 y + d_1 = 0 , \quad \text{and}$$

$$ax^2 + 2hxy + by^2 + 2p_2 x + 2q_2 y + d_2 = 0$$

be conic sections having the same quadratic terms. Let a line be drawn with direction δ, $|\delta| = 1$, and let $\mathbf{P}$ be the point on the line such that the product of its distances from the points of intersection of the line with the first conic section is equal to that for the second conic section. Then the locus of $\mathbf{P}$ is a straight line, independent of the choice of δ.

Solution: Let η end at $\mathbf{P}$. Then the line through $\mathbf{P}$ is $\xi = \eta + t\,\delta$. This line intersects the first conic section in two points corresponding to values h_1 and h_2 for which

$$h_1 h_2 = \frac{(\mathbf{A}_1 \eta, \eta)}{(\mathbf{Q}\delta, \delta)} .$$

Similarly for the second conic section, k_1 and k_2 are such that

$$k_1 k_2 = \frac{(\mathbf{A}_2 \eta, \eta)}{(\mathbf{Q}\delta, \delta)} .$$

Comparing $h_1 h_2$ and $k_1 k_2$ the proof is clear.

9.6 **Principal planes:** Find the principal planes of the surface

$$3x^2 + 2xy + 4yz + 2xz - 2x - 14y - 2z - 9 = 0$$

Solution 1: First we find a center if possible. A center is a solution of the equation

$$(x\ y\ z\ 1) \begin{pmatrix} 3 & 1 & 1 \\ 1 & 0 & 2 \\ 1 & 2 & 0 \\ -1 & -7 & -1 \end{pmatrix} = (0\ \ 0\ \ 0)$$

[see 8.13]. This equation gives

$$\begin{cases} 3x + y + z - 1 = 0 \\ x + 2z - 7 = 0 \\ x + 2y - 1 = 0 . \end{cases}$$

Solving we get

$$\left(-\frac{3}{2}, \frac{5}{4}, \frac{17}{4}\right) .$$

As the characteristic values of

$$\mathbf{Q} = \begin{pmatrix} 3 & 1 & 1 \\ 1 & 0 & 2 \\ 1 & 2 & 0 \end{pmatrix} ,$$

we get 1, 4, −2.

Now we obtain characteristic directions corresponding to these values. To the characteristic value 1 corresponds

$$(l\ m\ n) \begin{pmatrix} 2 & 1 & 1 \\ 1 & -1 & 2 \\ 1 & 2 & -1 \end{pmatrix} = (0\ 0\ 0) .$$

This equation gives

$$\begin{cases} 2l + m + n = 0 \\ l - m + 2n = 0 , \end{cases}$$

and we may choose the set (1,−1,−1).

To the characteristic value 4 corresponds

$$(l\ m\ n) \begin{pmatrix} -1 & 1 & 1 \\ 1 & -4 & 2 \\ 1 & 2 & -4 \end{pmatrix} = (0\ 0\ 0) ,$$

or

$$\begin{cases} -l + m + n = 0 \\ l - 4m + 2n = 0\ . \end{cases}$$

We choose the set (2,1,1).

Finally, to -2 corresponds

$$(l\ m\ n) \begin{pmatrix} 5 & 1 & 1 \\ 1 & 2 & 2 \\ 1 & 2 & 2 \end{pmatrix} = (0\ 0\ 0)\ .$$

Thus

$$\begin{cases} 5l + m + n = 0 \\ l + 2m + 2n = 0\ , \end{cases}$$

and we choose the set (0,1,–1).

Therefore the principal planes are

$$\left(x + \frac{3}{2}\right) - \left(y - \frac{5}{4}\right) - \left(z - \frac{17}{4}\right) = 0\ ,\quad 2\left(x + \frac{3}{2}\right) - \left(y - \frac{5}{4}\right) + \left(z - \frac{17}{4}\right) = 0\ ,\quad \left(y - \frac{5}{4}\right) - \left(z - \frac{17}{4}\right) = 0\ .$$

Solution 2: We find diametral planes corresponding to each characteristic direction [see 8.12]. For example corresponding to the direction (1,–1,–1) we get

$$(1\ -1\ -1\ 0) \begin{pmatrix} 3 & 1 & 1 & -1 \\ 1 & 0 & 2 & -7 \\ 1 & 2 & 0 & -1 \\ -1 & -7 & -1 & -9 \end{pmatrix} \begin{pmatrix} x \\ y \\ z \\ 1 \end{pmatrix} = (0)\ .$$

We leave the other cases as an exercise.

9.7 **Central quadric:** Change

$$3x^2 + 2xy + 2xz + 4yz - 2x - 14y - 2z - 9 = 0$$

to standard form.

Solution: Since the quadric has a center at $\left(-\frac{3}{2}, \frac{5}{4}, \frac{17}{4}\right)$, translation to this center eliminates first degree terms; the quadratic terms are then unchanged. Thus all we have to do is to compute the new constant which is obtained by the substitution

$$\begin{cases} x = x' - \dfrac{3}{2} \\ y = y' + \dfrac{5}{4} \\ z = z' + \dfrac{17}{4}\ . \end{cases}$$

Considering only the constant terms, we get

$$3\left(\frac{9}{4}\right) + 2\left(-\frac{3}{2}\right)\left(\frac{5}{4}\right) + 2\left(-\frac{3}{2}\right)\left(\frac{17}{4}\right) + 4\left(\frac{5}{4}\right)\left(\frac{17}{4}\right) - 2\left(-\frac{3}{2}\right) - 14\left(\frac{5}{4}\right) - 2\left(\frac{17}{4}\right) - 9 = 20\ .$$

Since 1, 4, and −2 are the characteristic values of the quadratic form of the quadric, one of the standard forms of the quadric is

$$x^2 + 4y^2 - 2z^2 + 20 = 0 \ .$$

9.8 **Quadric of rank 2:** Transform the equation

$$8x^2 - 4xy + 4xz - 2yz + 2x + 2y - 6z - 20 = 0$$

to standard form.

Solution: To locate a center we consider

$$(x\ y\ z\ 1) \begin{pmatrix} 8 & -2 & 2 \\ -2 & 0 & -1 \\ 2 & -1 & 0 \\ 1 & 1 & -3 \end{pmatrix} = (0\ 0\ 0)$$

or

$$\begin{cases} 8x - 2y + 2z + 1 = 0 \\ -2x \qquad - z + 1 = 0 \\ 2x - y \qquad - 3 = 0 \ . \end{cases}$$

Since the rank of the augmented matrix of this system of equations is greater than the rank of the coefficient matrix, there are no solutions. We therefore look for the principal planes. The characteristic values of the matrix

$$\mathbf{Q} = \begin{pmatrix} 8 & -2 & 2 \\ -2 & 0 & -1 \\ 2 & -1 & 0 \end{pmatrix}$$

are 0, −1, and 9, and the corresponding directions are (1,2,−2), (0,1,1), and (−4,1,−1). The latter directions give the principal planes

$$y + z + 2 = 0 \quad \text{and} \quad 4x - y + z = 0 \ ,$$

but the direction (1,2,−2) corresponding to the characteristic value 0 does not give a plane.

Since the rank of **Q** is 2, that of the matrix **A** of the quadric is 4, and the non-zero characteristic values have opposite signs, the surface is a hyperbolic paraboloid. The line of intersection of the two principal planes found above intersects the surface at a single point, which is the vertex of the surface. Since the point (0,−1,−1) is on this line, and it has direction (1,2,−2) the line may be written as

$$\begin{cases} x = t \\ y = -1 + 2t \\ z = -1 - 2t \ . \end{cases}$$

In the equation for the points of intersection, clearly $(\mathbf{Q}\delta,\ \delta) = 0$, and the equation becomes

$$18t - 18 = 0 \ ,$$

whence $t = 1$ and the vertex is (1,1,−3).

The change of coordinates (writing x, y, z in terms of x', y', z') is

$$\mathbf{S} = \begin{pmatrix} 0 & \frac{4\sqrt{2}}{6} & \frac{1}{3} & 1 \\ \frac{\sqrt{2}}{2} & \frac{-\sqrt{2}}{6} & \frac{2}{3} & 1 \\ \frac{\sqrt{2}}{2} & \frac{\sqrt{2}}{6} & -\frac{2}{3} & -3 \\ 0 & 0 & 0 & 1 \end{pmatrix},$$

and to compute $\mathbf{B} = \mathbf{S}' \mathbf{A} \mathbf{S}$, we observe that only a_{41}, a_{42}, and a_{43} of the matrix $\mathbf{B}$ have to be computed. We get

$$(1 \quad 1 \quad -3 \quad 1)\ \mathbf{A}\,\mathbf{S} = (0 \quad 0 \quad 3 \quad 0)\,,$$

whence

$$\mathbf{B} = \begin{pmatrix} -1 & 0 & 0 & 0 \\ 0 & 9 & 0 & 0 \\ 0 & 0 & 0 & 3 \\ 0 & 0 & 3 & 0 \end{pmatrix}$$

and the transformed equation has the form

$$-x^2 + 9y^2 + 6z = 0\,.$$

9.9 **Quadric of rank 1:** Transform the equation

$$x^2 + 4y^2 + 4z^2 + 4xy - 4xz - 8yz + 2x + 8y + 7 = 0$$

to standard form.

Solution: Since there is no center we look for principal planes. The characteristic values of

$$\mathbf{Q} = \begin{pmatrix} 1 & 2 & -2 \\ 2 & 4 & -4 \\ -2 & -4 & 4 \end{pmatrix}$$

are 0, 0, and 9. The direction corresponding to 9 is (1,2,–2) and the corresponding principal plane is $x + 2y - 2z + 1 = 0$. Since the surface is a parabolic cylinder, this plane cuts the surface in a line of vertices, which is the solution of

$$\begin{cases} x^2 + 4y^2 + 4z^2 + 4xy - 4xz - 8yz + 2x + 8y + 7 = 0 \\ x + 2y - 2z + 1 = 0\,. \end{cases}$$

Since the standard equation of the parabolic cylinder has only one second degree term, the quadratic terms of the first equation are proportional to the square of the linear terms in the second. Thus

$$(x + 2y - 2z)^2 = 1$$

and the line of vertices is the line of intersection of the planes

$$x + 2y - 2z + 1 = 0 \quad \text{and} \quad x + 4y + 4 = 0\,.$$

The direction of this line is (4,–1,1). For the new coordinate system we may choose (1,2,–2) as the x-direction, (4,–1,1) as the z-direction, and the third orthogonal direction is then (0,1,1). Choosing the point

$\left(0,-1,-\frac{1}{2}\right)$ on the line of vertices we get for the change of coordinates

$$\mathbf{S} = \begin{pmatrix} \frac{1}{3} & 0 & \frac{4\sqrt{2}}{6} & 0 \\ \frac{2}{3} & \frac{\sqrt{2}}{2} & \frac{-\sqrt{2}}{6} & -1 \\ \frac{-2}{3} & \frac{\sqrt{2}}{2} & \frac{\sqrt{2}}{6} & -\frac{1}{2} \\ 0 & 0 & 0 & 1 \end{pmatrix} .$$

As in 9.8, the transformed equation is of the form

$$9x^2 + 4\sqrt{2}\, y = 0 .$$

9.10 **Axis of rotation:** Find the axis of the rotation with matrix

$$\begin{pmatrix} \frac{6}{7} & -\frac{3}{7} & \frac{2}{7} \\ \frac{2}{7} & \frac{6}{7} & \frac{3}{7} \\ -\frac{3}{7} & -\frac{2}{7} & \frac{6}{7} \end{pmatrix} .$$

Solution: Since the characteristic values of a rotation are 1 and λ, $\bar{\lambda}$ where $|\lambda| = 1$, the direction corresponding to 1 is invariant under the rotation and is called the axis of rotation. The equation of this line is

$$(x\ y\ z) \begin{pmatrix} \frac{6}{7} & -\frac{3}{7} & \frac{2}{7} \\ \frac{2}{7} & \frac{6}{7} & \frac{3}{7} \\ \frac{-3}{7} & \frac{-2}{7} & \frac{6}{7} \end{pmatrix} = (x\ y\ z) ,$$

or

$$\begin{cases} 6x + 2y - 3z = 7x \\ -3x + 6y - 2z = 7y . \end{cases}$$

We leave it as an exercise to change this to parametric form.

9.11 **Identification of a quadric:** We wish to consider the cases in which it is impractical to find the characteristic values of the matrix **Q** of the quadric. These are the cases i-vi of 8.9 in which it may happen that none of the characteristic values is rational.

We discuss the equation

$$x^2 + 3z^2 - 2xz + 2yz + 2y + 6z + 9 = 0 .$$

Since the matrix

$$\mathbf{Q} = \begin{pmatrix} 1 & 0 & -1 \\ 0 & 0 & 1 \\ -1 & 1 & 3 \end{pmatrix} ,$$

the characteristic equation is

$$-\lambda^3 + 4\lambda^2 - 3\lambda - 1 = 0 .$$

The rank of **Q** is 3, hence the quadric is one of the cases mentioned above. The determinant of the matrix

$$\mathbf{A} = \begin{pmatrix} 1 & 0 & -1 & 0 \\ 0 & 0 & 1 & 1 \\ -1 & 1 & 3 & 3 \\ 0 & 1 & 3 & 9 \end{pmatrix}$$

is not zero, thus cases iii and vi may be eliminated. Since det **A** is negative, we may eliminate ii and v. Finally since det **Q** = −1, the product of the characteristic roots is negative, while their sum is positive, hence there must be two positive roots and one negative root. Thus the surface is a hyperboloid of two sheets.

The signs of the characteristic roots may always be determined by the use of the Descartes rule of signs.

9.12 **Rulings:** Find the rulings of

$$\frac{x^2}{a^2} + \frac{y^2}{b^2} - \frac{z^2}{c^2} - 1 = 0$$

at (x_o, y_o, z_o) of the quadric.

Solution: By 8, 15, (λ, μ, ν), the direction of the rulings is obtained from

$$(1) \qquad \begin{cases} \dfrac{x_o}{a^2}\lambda + \dfrac{y_o}{b^2}\mu - \dfrac{z_o}{c^2}\nu = 0 \\[2ex] \dfrac{1}{a^2}\lambda^2 + \dfrac{1}{b^2}\mu^2 - \dfrac{1}{c^2}\nu^2 = 0 \end{cases}$$

Let $\dfrac{\lambda}{\nu} = l$ and $\dfrac{\mu}{\nu} = m$. Solving (1) we get

$$l = \frac{a^2b^2z_ox_o \pm a^3cby_o}{a^2c^2y_o^{\,2} + b^2c^2x_o^{\,2}}, \qquad m = \frac{a^2b^2z_oy_o \pm b^3cax_o}{a^2c^2y_o^{\,2} + b^2c^2x_o^{\,2}},$$

and the rulings can be written as

$$\begin{cases} x = x_o + (abz_ox_o \pm a^2cy_o)t \\ y = y_o + (abz_oy_o \pm b^2cx_o)t \\ z = z_o + ab(z_o^{\,2} + c^2)t \end{cases} .$$

9.13 **Locus problems:**

Problem 1: Determine an equation of the locus of a point equidistant from the lines

$$\begin{cases} x = 2s \\ y = s \\ z = 0 \end{cases} \quad \text{and} \quad \begin{cases} x = t \\ y = 0 \\ z = 1 + t \end{cases} .$$

Solution: For a point $P : (x,y,z)$ of the locus there are points P_1 and P_2 respectively on the given lines such that PP_1 is perpendicular to the first line, PP_2 is perpendicular to the second, and PP_1 and PP_2 have the same length. Let s and t be values of the parameters corresponding to P_1 and P_2 respectively. Then using inner product and norm we get

$$\begin{cases} 2(x-2s)+(y-s)=0 \\ (x-t)+(z-1-t)=0 \\ (x-2s)^2+(y-s)^2+z^2=(x-t)^2+y^2+(z-1-t)^2 \ . \end{cases}$$

Finding s and t in terms of x,y,z from the first two equations, and substituting in the third we obtain the desired equation.

Problem 2: Find the equation of the circular cylinder having radius 2, and axis the line through the origin with direction (1,2,3).

Solution: Let $P = (x,y,z)$ be a point of the cylinder, $H = (t,2t,3t)$ a point of the axis such that PH is perpendicular to the axis. As above, again using norm and inner product, we have

$$\begin{cases} (x-t)+2(y-2t)+3(z-3t)=0 \\ (x-t)^2+(y-2t)^2+(z-3t)^2=4 \ . \end{cases}$$

Eliminating t we get

$$13x^2+10y^2+5z^2-4xy-6xz-12yz-56=0 \ .$$

Problem 3: Find an equation of the surface generated when the line

$$\begin{cases} x=1-s \\ y=3s \\ z=-3s \end{cases}$$

is rotated about the line

$$\begin{cases} x=2t \\ y=2+t \\ z=1-2t \ . \end{cases}$$

Solution: If $P = (x,y,z)$ is a point of the surface, let the plane through P perpendicular to the second line cut it at $P_1 = (2t,\ 2+t,\ 1-2t)$, and cut the first line at $P_2 = (1-s,\ 3s,\ -3s)$. We observe that PP_1 and PP_2 are both perpendicular to the second line, while PP_1 and P_1P_2 have the same length. As before, we get

$$\begin{cases} 2(x-2t)+(y-2-t)-2(z-1+2t)=0 \\ 2(x-1+s)+(y-3s)-2(z+3s)=0 \\ (x-2t)^2+(y-2-t)^2+(z-1+2t)^2=(1-s-2t)^2+(3s-2-t)^2+(-3s-1+2t)^2 \ . \end{cases}$$

The desired equation is found as before.

9.14 **Curves in space:** As a line in space is usually defined by two planes containing it, so we may locate a curve in space by means of the equations of two surfaces containing it. A curve may also be described by a set of parametric equations of the form

$$\begin{cases} x=f_1(t) \\ y=f_2(t) \\ z=f_3(t) \ . \end{cases}$$

We consider certain problems concerning curves, which are related to the surfaces discussed. We observe that if $f_1(x,y,z) = 0$ and $f_2(x,y,z) = 0$ are equations of any two surfaces, then $af_1 + bf_2 = 0$, where a and b are any polynomials in x, y, z, is the equation of a surface containing all points of intersection of the two given surfaces.

Problem 1: Show that the curve

$$\begin{cases} 3x^2 + y^2 + y + x = 0 \\ 6x^2 + 2y^2 + z + 2x = 0 \end{cases}$$

lies in a plane.

Solution: We see that

$$(6x^2 + 2y^2 + z + 2x) - 2(3x^2 + y^2 + y + x) = 6x^2 + 2y^2 + z + 2x - 6x^2 - 2y^2 - 2y - 2x = z - 2y = 0$$

is the equation of a plane containing the curve.

Problem 2: Find the projection on xz-plane of the curve

$$\begin{cases} xy + xz + yz = 2 \\ 4xy - 2xz + 3yz + y - 2z = 6\ . \end{cases}$$

Solution: From the first equation

$$y = \frac{2 - xz}{x + z}\ .$$

This may be substituted in the second equation to give the equation of a cylindrical surface whose generator is perpendicular to the xz-plane. This equation, with the equation $y = 0$, gives the projection.

9.15 **Pole and polar:** For a given quadric in the space there are some interesting results obtainable from the relation of pole and polar. We wish to consider a few of them through the following problems.

Problem 1: Find the pole of $4x + y + 2z + 2 = 0$ with respect to

$$x^2 + 4xy + 4yz - 8z = 0\ .$$

Solution: We want to find a point (x_o, y_o, z_o) such that

$$(x_o\ y_o\ z_o\ 1)\begin{pmatrix} 1 & 2 & 0 & 0 \\ 2 & 0 & 2 & 0 \\ 0 & 2 & 0 & -4 \\ 0 & 0 & -4 & 0 \end{pmatrix}\begin{pmatrix} x \\ y \\ z \\ 1 \end{pmatrix} \equiv k(4x + y + 2z + 2)\ ,$$

k is a constant. This gives

$$\begin{cases} x_o + 2y_o = 4k \\ 2x_o + 2z_o = k \\ 2y_o - 4 = 2k \\ \quad - 4z_o = 2k\ . \end{cases}$$

The solution of this system gives the point (4,6,–2).

Problem 2: As a point **P** varies on the line

$$\begin{cases} x = 2t \\ y = 1 - t \\ z = 2 + t\ , \end{cases}$$

show that the polar of P with respect to

$$2xy - 4yz - 6xz + 8x - 1 = 0$$

revolves around a fixed line.

Solution:

$$(2t \;\; 1-t \;\; 2+t \;\; 1)\begin{pmatrix} 0 & 1 & -3 & 4 \\ 1 & 0 & -2 & 0 \\ -3 & -2 & 0 & 0 \\ 4 & 0 & 0 & -1 \end{pmatrix}\begin{pmatrix} x \\ y \\ z \\ 1 \end{pmatrix} = (0)$$

gives

$$(-1-4t)\, x - 4y + (-2-4t)\, z - 1 + 8t = 0$$

or

$$-x - 4y - 2z - 1 + t\,(-4x - 4z + 8) = 0 \,.$$

For any t, this a plane through the line

$$\begin{cases} -x - 4y - 2z - 1 = 0 \\ -4x - 4z + 8 = 0 \,. \end{cases}$$

EXERCISES 9

1. Describe the projection of the space on the plane $z = 3x - 2y$.
2. Given two lines

$$(1) \quad \begin{cases} x = x_1 + tl_1 \\ y = y_1 + tm_1 \\ z = z_1 + tn_1 \end{cases} \qquad (2) \quad \begin{cases} x = x_2 + sl_2 \\ y = y_2 + sm_2 \\ z = z_2 + sn_2 \,, \end{cases}$$

find:

(i) the condition that (1) and (2) are intersecting or parallel (coplanar),
(ii) the line segment AB such that A is on (1), B is on (2), and AB is perpendicular to both lines.

3. Determine the relationship of the planes

$$\text{(i)} \quad \begin{cases} x + y = 2 \\ y + z = 2 \\ x + z = 2 \,, \end{cases} \qquad \text{(ii)} \quad \begin{cases} x + y = 1 \\ x + z = 2 \\ x + y + z = 4 \,, \end{cases}$$

$$\text{(iii)} \quad \begin{cases} x + y - 2z = -1 \\ 2x - y + z = 4 \\ 4x + y - 3z = 5 \,, \end{cases} \qquad \text{(iv)} \quad \begin{cases} x + y - 2z = -1 \\ 2x - y + z = 4 \\ x + 4y - 7z = -7 \,. \end{cases}$$

4. Find the distance between a diagonal of a cube of edge a and the edges which do not meet that diagonal.
5. In a circle of radius r two perpendicular chords AB and CD intersect at P. Show that

$$PA^2 + PB^2 + PC^2 + PD^2 = 4r^2 \,.$$

6. Find the principal planes of

(i) $x^2 + y^2 - 7z^2 - 2xy - 4xz - 4yz + 8y + 14z - 6 = 0$,
(ii) $x^2 - 3y^2 + 3z^2 + 8yz + 6x + 29 = 0$,
(iii) $x^2 - z^2 + 2xy + 4yz - 6x - 8y - 2z + 9 = 0$,
(iv) $4y^2 - 2xy - xz + 2yz + x + y - 3z - 10 = 0$,
(v) $5x^2 + 3y^2 + 3z^2 + 2xy - 2xz - 2yz - 12 = 0$.

7. Change the quadrics of exercise 6 to standard form.
8. Find the matrix and the axis of each of the rotations of exercise 7.
9. Find the rulings of

(i) $x^2 + y^2 - 2z^2 = 1$ at any point on it,
(ii) $\frac{x^2}{25} - y^2 + 2z = 0$ with direction (5,–1,–6),
(iii) $\frac{x^2}{16} + \frac{y^2}{4} - z^2 = 1$ with direction (16,–6,5).

10. Show that the rulings in 9.12 can be written with direction numbers

$$a, \pm b, \frac{-x_o}{a} \pm \frac{y_o}{b} .$$

11. Find the rulings of

$$\frac{x^2}{a^2} - \frac{y^2}{b^2} + 2z = 0$$

at (x_o, y_o, z_o).

12. Find the radical plane of

$$x^2 + y^2 + z^2 - 2x - 2y - 2z = 0 \text{ and}$$
$$(x-7)^2 + (y-2)^2 + (z-1)^2 = 25 .$$

13. Find the polar of the given point with respect to the given conic section or quadric

(i) (2,7) ; $3x^2 - 2xy + x - 2 = 0$,
(ii) (3,0,–4) ; $xy - xz + 2yz - 5 = 0$
(iii) (1,1,1) ; $3x^2 - y^2 + 2z^2 + 4xy - 10xz - 2y + 6z = 0$.

14. Find the pole of each line or plane with respect to the given conic section or quadric

(i) $x - y = 0$; $x^2 - 3xy + y = 2$,
(ii) $5x - 4y - 3z = 15$; $5x^2 - y^2 - 6z^2 - 30 = 0$,
(iii) $x - y = 1$; $xy - yz + zx = 1$,
(iv) $ax + by + cz + d = 0$; $(\mathbf{A}\xi, \xi) = 0$, where $\mathbf{A}$ is a symmetric matrix.

15. Change to diagonal form

(i) $\begin{pmatrix} 3 & 1-i \\ 1+i & 2 \end{pmatrix}$, (ii) $\begin{pmatrix} 1 & i \\ -i & 1 \end{pmatrix}$.

16. Identify the following quadrics:

(i) $x^2 - 2y^2 + 4xy - 2xz + 6x - 4z - 12 = 0$,
(ii) $3x^2 + y^2 - z^2 + 2xy - 4xz + 10y - 4 = 0$,
(iii) $x^2 + 4y^2 + 4z^2 + 4xy + 4xz + 8yz + 8x - 4y + 2 = 0$,
(iv) $-3x^2 + 24y^2 - z^2 + 12xy - 6xz - 12yz + 8x - 10y + 2z = 0$,
(v) $2x^2 + 4y^2 - z^2 + 6xy - 8yz + 10x + 2y - 4 = 0$.

17. Find the equation of the surface generated by revolving the line

$$x = 0, \quad y = 1$$

about the line

$$x = \frac{y}{2} = \frac{z}{3} .$$

18. Find the equation of the surface generated by revolving the line

$$\begin{cases} x = 2 - t \\ y = 3 - 2t \\ z = -1 + t \end{cases}$$

about the line

$$\begin{cases} x = 2 - 2s \\ y = 3 + s \\ z = -1 \end{cases} .$$

19. Find the locus of a point equidistant from the lines

$$x = 2y = 0, \; z = 0 \quad \text{and} \quad 2x - y = 0, \; z = 0 ,$$

and identify the surface.

20. Find the projection on the coordinate planes of the curve

$$\begin{cases} x^2 + 3yz - 6y + z = 4 \\ x^2 + y - 2z = 2 \end{cases} .$$

21. Find the polar of the given point with respect to the given quadric:

(i) $(1,2,3)$; $x^2 - 3z^2 - 2xz - y - 1 = 0$,
(ii) $(-1,-3,0)$; $3x^2 - 7y^2 + yz + 4x - 2 = 0$.

22. Find the pole of the given plane with respect to the given quadric:

(i) $x - 2y + z = 0$; $3x^2 - z^2 - xy - yz - x - 3 = 0$,
(ii) $2x - z = 1$; $x^2 + y^2 + z^2 - 2xy - 6xz - 7x - 7 = 0$.

23. Find the polar of the line

$$x = \frac{y-1}{2} = \frac{z-3}{3}$$

with respect to

$$x^2 - 3y^2 - z^2 - xy - yz - xz - 2x + z - 1 = 0 .$$

PART III

10. SOME ALGEBRAIC STRUCTURES

Introduction: We are now interested in introducing the reader to the modern methods and results in the theory of linear spaces. From the concrete study of Euclidean three-dimensional space we have generalized to complex n-dimensional spaces. Now we want to take the third step in the direction of abstraction and generalization. We consider first some basic algebraic structures and the general notion of vector space, then some of the more recent results in the case of real and complex unitary spaces. We will not go into too much detail, as there are many adequate treatments of the material on an advanced level. We want merely to acquaint the student with the style and content of modern abstract algebra.

10.1 **Definition:** Let us accept intuitively the idea of a *set* S as a collection of *elements* $\{a,b,c, \ldots\}$. If a is in S, or a is an *element* of S, we write $a \epsilon$ S.

A *single-valued binary operation* "ω" on S is a rule which assigns to each ordered pair of elements a and b of S a unique object $p = a \omega b$, which may or may not be in S. If for any two elements a and b of S, $p = a \omega b \epsilon$ S, then S is said to be *closed* under "ω".

A set S is said to be *commutative* under a single-valued binary operation "ω" when for any two objects $a \epsilon$ S and $b \epsilon$ S we have $a \omega b = b \omega a$.

A set S is called *associative* under "ω" if for any three elements a,b,c of S we have $a \omega (b \omega c) = (a \omega b) \omega c$.

10.2 **Groups:** A set G is called a *group* under a single-valued binary operation "ω" if:

I G is closed under "ω",
II G is associative under "ω",
III There is an element e in G, called the *identity*, such that for any $a \epsilon$ G we have $e \omega a = a \omega e = a$,
IV For any $a \epsilon$ G there is an $a' \epsilon$ G, called the *inverse* of a, such that $a' \omega a = a \omega a' = e$.

We denote a' by a^{-1}.

Now if in addition G is commutative under "ω" the group is called *commutative* (*Abelian*). We shall write ab for $a \omega b$.

10.3 **Theorem:** Let $a,b \epsilon$ G, and G be a group. Then $xa = b$ and $ay = b$ have unique solutions $x = ba^{-1}$ and $y = a^{-1}b$. This immediately implies that if $ca = da$, then $c = d$.

Proof: We observe that

$$x = xe = xaa^{-1} = ba^{-1} ;$$

also

$$y = ey = a^{-1}ay = a^{-1}b .$$

Conversely we see that

$$(ba^{-1})a = b \quad \text{and} \quad a(a^{-1}b) = b .$$

10.4 **Corollary:** A group G has a unique identity, and each element has a unique inverse.

Proof: Clearly $ex = e$ implies $x = e$, and $ay = e$ implies $y = a^{-1}$. This proves the corollary.

10.5 **Fields:** A *field* F is a set which is closed under two single-valued binary operations, say "+" and ".", called addition and multiplication, such that

I F is a commutative group under "+" with the identity 0.
II The set consisting of all elements of F except 0 is a commutative group under multiplication.
III Multiplication is *distributive* over addition.

I.e. for $a, b, c \in F$ we have

$$a(b+c) = ab + ac .$$

10.6 **Examples:**

I Let G = {all integers; positive, negative, zero}, and let the binary operation be addition. Then G is a commutative group.

II Let G = {all rotations in three dimensional real space}, and let the binary operation be the multiplication of these transformations. Then G is a group, but not commutative.

III Let F = {all rational numbers}, and the two operations be addition and multiplication. Then F is a field.

IV Put all the integers into five categories and respectively call them:

$$e = \{\text{any multiple of } 5\} ,$$
$$a = \{(\text{any multiple of } 5) + 1\} ,$$
$$b = \{(\text{any multiple of } 5) + 2\} ,$$
$$c = \{(\text{any multiple of } 5) + 3\} ,$$
$$d = \{(\text{any multiple of } 5) + 4\} .$$

This clearly exhausts all the integers. Let $F = \{e, a, b, c, d\}$. Then F is a field under addition and multiplication, where e is the identity of the addition.

10.7 **Vector spaces:** A *vector space* V over a field F is a set of elements, called *vectors*, α, β, . . . such that:

I V is a commutative group under addition.

II To any $a \in F$ and $\alpha \in V$ corresponds a vector $\beta = a\alpha \in V$

III For $a, b \in F$ and $\alpha, \beta \in V$,

(1) $a(\alpha+\beta) = a\alpha + a\beta$, and

(2) $(a + b)\alpha = a\alpha + b\alpha$,

(3) $(ab)\alpha = a(b\alpha)$,

(4) $1\alpha = \alpha$.

The elements of F are called scalars.

The identity of the commutative group V is called the *zero vector*. We denote it by 0. It is easily seen that the following relations between the zero of F, i.e., 0, and the zero vector are true.

$$0\alpha = 0 \quad \text{for} \quad \alpha \in V ,$$

$$c0 = 0 \quad \text{for} \quad c \in F .$$

Also it is interesting to observe that $(-1)\alpha$ is the inverse of $\alpha \in V$. This can be proved as follows:

$$\alpha + (-1)\alpha = (1)\alpha + (-1)\alpha = [(1) + (-1)]\alpha = 0\alpha = 0 .$$

10.8 **Subspaces:** A subset S of a vector space V over the field F is called a *subspace* if S is a vector space over the same field with respect to the same addition and multiplication by a scalar as in V. Note that the set S is contained in V. We write this as $S \subset V$. It is easily proved that a subset S of V is a subspace if S is closed under addition and multiplication by a scalar in V. We leave the proof to the reader.

10.9 **Examples of vector spaces:** In Chapter 1 we presented the simplest example of a vector space over the field of real numbers. We have also introduced vector spaces over the field of complex numbers. Here we would like to give some examples which are more general.

I The set of all polynomials of degree n with real coefficients is a vector space over the field of real numbers. We verify this by examining these polynomials according to 10.7. Let $a_o + a_1x + \ldots + a_nx^n$ be a polynomial of degree n. Then we easily see that the set of such polynomials is a commutative group under addition. Clearly $a(a_o + a_1x + \ldots + a_nx^n)$ is also a polynomial of degree n.

We also see that

$$a(a_o + a_1x + \ldots + a_nx^n) + b(b_o + b_1x + \ldots + b_nx^n)$$

$$= (aa_o + bb_o) + (aa_1 + bb_1)x + \ldots + (aa_n + bb_n)x^n .$$

Similarly the rest of the properties of a vector space can be checked.

II The set of all real continuous functions with domain [0,1]. This can easily be verified. We leave it to the reader as an exercise.

10.10 **Linear independence:** A set of vectors $\{\alpha_1, \ldots, \alpha_n\}$ in a vector space V over a field F is called *linearly independent* if for $c_1, \ldots, c_n \in F$,

$$c_1\alpha_1 + \ldots + c_n\alpha_n = 0$$

if and only if $c_1 = \ldots = c_n = 0$; otherwise $\{\alpha_1, \ldots, \alpha_n\}$ is called *linearly dependent.*

Clearly any set of vectors containing 0 is linearly dependent.
Here theorem 6.4 is true and proved in exactly the same way.

10.11 **Base:** A set $\{\alpha_i\}$ of vectors of a vector space V over a field F is called a *base* if:

(1) Any finite subset of $\{\alpha_i\}$ is linearly independent
(2) Any vector of V can be written as a finite linear combination of elements of $\{\alpha_i\}$.

We say $\{\alpha_i\}$ generates the space.

It is left to the reader to show that any vector in V can be written in terms of the elements of a base in one and only one way. If V has a finite base we call it a *finite dimensional* vector space.

The theorems 6.4 and 6.6 are true for any vector space V over a field F. Since the proof in general is exactly as is done in 6.6 we will not repeat it here.

10.12 **Theorem:** Any two bases for a finite dimensional vector space V over a field F have the same number of elements.

Proof: Let $\{\alpha_1, \ldots, \alpha_n\}$ be a base for V.

Suppose $\{\beta_1, \ldots, \beta_m\}$ is another base for V.

Since $\{\alpha_1, \ldots, \alpha_n\}$ is a base for V every β_i, $i = 1, \ldots, m$ is a linear combination of α's. If $m > n$, then by 6.6 $\beta_1, \ldots, \beta_m$ is linearly dependent. Therefore $m \leq n$. On the other hand, since $\{\beta_1, \ldots, \beta_m\}$ is a base for V, every α_i, $i = 1, \ldots, n$, is a linear combination of β's. If $n > m$ then by 6.6 $\{\alpha_1, \ldots, \alpha_n\}$ is linearly dependent. Therefore $m = n$.

Now we define the number of elements in a base for V to be the *dimension* of V.

10.13 **Corollary:** Let $\{\alpha_1, \ldots, \alpha_k\}$ be a set of linearly independent vectors in V, a vector space over a field F. Clearly $k \leq n$, the dimension of V. Then there exists a base in V such that $\{\alpha_1, \ldots, \alpha_k\}$ is a part of it.

The proof is closely related to 6.6 and is left to the reader.

10.14 **Theorem:** Let S and T be two subspaces of V, a vector space over a field F. Then the common part of S and T, denoted by $S \cap T$, is also a subspace of V. But the set of all vectors of S and T is not necessarily a subspace.

Proof: Let $\alpha \in S \cap T$. This means that $\alpha \in S$ and also $\alpha \in T$. Since S and T are subspaces, by 10.8, for any scalar a we have $a\ \alpha \in S$ also $a\ \alpha \in T$. Therefore $a\ \alpha \in S \cap T$. Now let $\alpha \in S \cap T$ and $\beta \in S \cap T$. This implies that $\alpha \in S$ and $\alpha \in T$ and also $\beta \in S$ and $\beta \in T$. Again by 10.8 we have $\alpha + \beta \in S$ and $\alpha + \beta \in T$ which means $\alpha + \beta \in S \cap T$.

For the second part of the theorem we give a counter example. Let S be all the vectors on the x-axis and T be all the vectors on the y-axis of a three dimensional Euclidean space. Clearly the sum $\alpha + \beta$, $\alpha \in S$ and $\beta \in T$ where α and β are non-zero vectors, is neither in S nor in T, but $\alpha + \beta$ is in the plane $z = 0$.

Now for two subspaces S and T we define S + T to be the set of all vectors of the form $\alpha + \beta$ where $\alpha \epsilon S$ and $\beta \epsilon T$.

The reader may show that S + T is a subspace such that S and T both are contained in it, that is,

$$S \subset S + T \text{ and } T \subset S + T .$$

10.15 **Theorem:** Let dim S denote the dimension of the subspace S. Then

$$\dim S + \dim T = \dim (S + T) + \dim (S \cap T)$$

Proof: Let $\{\alpha_1, \ldots, \alpha_k\}$ be a base in $S \cap T$.

This means $\dim (S \cap T) = k$. By 10.13, since $S \cap T \subset S$ and $S \cap T \subset T$, $\{\alpha_1, \ldots, \alpha_k\}$ is a part of a base $\{\alpha_1, \ldots, \alpha_k, \xi_1, \ldots, \xi_s\}$ for S, and part of a base $\{\alpha_1, \ldots, \alpha_k, \eta_1, \ldots, \eta_t\}$ for T. This means

$$\dim S = k + s \text{ and } \dim T = k + t .$$

Now clearly any linear combination of the α's, ξ's, and η's is in S + T. Let

$$(1) \qquad a_1 \alpha_1 + \ldots + a_k \alpha_k + b_1 \xi_1 + \ldots + b_s \xi_s + c_1 \eta_1 + \ldots + c_t \eta_t = 0 .$$

We see that no non-zero linear combination of the ξ's is in T. For if some $b_1 \xi_1 + \ldots b_s \xi_s$ were in T, it would also be in S, hence in $S \cap T$, and would be a linear combination of the α's. This contradicts the assumption that $\{\alpha_1, \ldots, \alpha_k, \xi_1, \ldots, \xi_s\}$ is linearly independent. Similarly no non-zero linear combination of the η's is in S.

But from (1),

$$b_1 \xi_1 + \ldots + b_s \xi_s = - a_1 \alpha_1 - \ldots - a_k \alpha_k - c_1 \eta_1 - \ldots - c_t \eta_t \epsilon T .$$

Thus $b_1 \xi_1 + \ldots + b_s \xi_s = 0$, and $b_1 = \ldots = b_s = 0$. Similarly

$$c_1 \eta_1 + \ldots + c_t \eta_t = - a_1 \alpha_1 - \ldots - a_k \alpha_k - b_1 \xi_1 - \ldots - b_s \xi_s ,$$

which implies $c_1 = \ldots = c_t = 0$. Thus (1) implies

$$a_1 \alpha_1 + \ldots + a_k \alpha_k = 0 .$$

and since the α's are linearly independent

$$a_1 = \ldots = a_k = 0 .$$

Therefore $\{\alpha_1, \ldots, \alpha_k, \xi_1, \ldots, \xi_s, \eta_1, \ldots, \eta_t\}$ is a base for S + T and

$$\dim (S + T) = k + s + t .$$

Consequently

$$\dim S + \dim T = \dim (S + T) + \dim (S \cap T) .$$

10.16 **Unitary spaces:** Let the field from which the scalars are chosen be the field of complex numbers. We define an *inner product* of two vectors ξ and η of V, a vector space over the complex numbers, to be a function f of the ordered pair (ξ, η) such that

(1) $f(\xi,\eta) = \overline{f(\eta,\xi)}$, where

$\overline{f(\eta,\xi)}$ is the complex conjugate of $f(\eta,\xi)$,

(2) $f(\xi,\xi) \geq 0$, $f(\xi,\xi) = 0$ if and only if $\xi = 0$.

(3) $f(a_1\xi_1 + a_2\xi_2, \eta) = a_1 f(\xi_1,\eta) + a_2 f(\xi_2,\eta)$, for $\xi_1, \xi_2, \eta \epsilon V$.

If the field is the field of real numbers, we use the same definition and note that the inner product is then commutative.

A vector space V over the field of real or complex numbers is called *unitary* if an inner product in V is defined.

Examples of inner product were given in 1.9, 6.8. We would like to give an example more general than those.

Let V be the vector space of all polynomials of degree n, with complex coefficients, in the real variable x over the region $[-1, +1]$. Then for $P_1 = P_1(x)$ and $P_2 = P_2(x)$, two polynomials, we define the inner product to be

$$f(P_1, P_2) = \int_{-1}^{+1} P_1(x)\, \overline{P_2(x)}\, dx .$$

The norm of a vector $\xi \epsilon V$ is defined to be $|\xi| = [f(\xi,\eta)]^{\frac{1}{2}}$. For example, we define the norm of the polynomial P to be

$$|P| = \left[\int_{-1}^{+1} P(x)\, \overline{P(x)}\, dx\right]^{\frac{1}{2}} .$$

For convenience we shall use the symbol (ξ,η) for the inner product of ξ and η.

10.17 **Theorem:** Let ξ and η be two vectors in a vector space V over the complex field, and c a complex number. Then

(1) $|c\,\xi| = |c|\,|\xi|$,

(2) $|(\xi,\eta)| \leq |\xi|\,|\eta|$ (Schwarz inequality),

(3) $|\xi + \eta| \leq |\xi| + |\eta|$ (triangle inequality).

Proof:

$$|c\,\xi|^2 = (c\,\xi,\, c\,\xi) = c\,\bar{c}\,(\xi,\, \xi) = |c|^2\,|\xi|^2 .$$

Therefore $|c\,\xi| = |c|\,|\xi|$. This proves (1).

Now let α, $\beta \epsilon V$ be such that $|\alpha| = |\beta| = 1$. Then

$$|\alpha + \beta|^2 = (\alpha + \beta,\ \alpha + \beta) = 2 + 2R(\alpha, \beta) \geq 0 ,$$

where $R(\alpha,\beta)$ means the real part of (α,β). This implies $R(\alpha,\beta) \geq -1$. Also

$$|\alpha - \beta|^2 = (\alpha-\beta,\ \alpha-\beta) = 2 - 2R(\alpha,\beta) \geq 0.$$ That is

$$R(\alpha,\beta) \leq 1 .$$

Therefore $|R(\alpha,\beta)| \leq 1$. Let $\alpha = \dfrac{\xi}{|\xi|}$ and $\beta = \dfrac{\eta}{|\eta|}$, $\xi, \eta \neq 0$. Then

(4) $|R(\xi,\eta)| \leq |\xi|\,|\eta|$

Now let a be a complex number for which $|a| = 1$.

Applying (4) to $(a\,\xi,\eta)$ we get

$$|R(a\,\xi,\eta)| \leq |a\,\xi|\,|\eta| = |a|\,|\xi|\,|\eta| = |\xi|\,|\eta|$$

or

(5) $|R\,a(\xi,\eta)| \leq |\xi|\,|\eta|$.

Now let $a = \cos\theta + i\sin\theta$, and $(\xi,\eta) = |(\xi,\eta)|\,(\cos\phi + i\sin\phi)$, the polar representations of a and (ξ,η). Then by the De Moivre Theorem

$$a(\xi,\eta) = |(\xi,\eta)|\,(\cos(\theta + \phi) + i\sin(\theta + \phi)) .$$

Since (5) is true for all a with $|a| = 1$, let $\theta = -\phi$. Then we have

$$a(\xi,\eta) = |(\xi,\eta)|\ , \quad \text{and (5) will be}$$

$$|(\xi,\eta)| \leq |\xi|\ |\eta|\ .$$

Now for (3) we see that

$$|\xi + \eta|^2 = |\xi|^2 + |\eta|^2 + 2R(\xi,\eta)\ .$$

But we have proved that $R(\alpha,\beta) \leq 1$ for vectors α and β with $|\alpha| = |\beta| = 1$, and choosing $\alpha = \dfrac{\xi}{|\xi|}$, and $\beta = \dfrac{\eta}{|\eta|}$, $\xi, \eta \neq 0$ we get

$$R(\xi,\eta) \leq |\xi|\ |\eta|\ .$$

Therefore

$$|\xi + \eta|^2 \leq |\xi|^2 + |\eta|^2 + 2|\xi|\ |\eta| = (|\xi| + |\eta|)^2\ ,$$

and consequently

$$|\xi + \eta| \leq |\xi| + |\eta|\ .$$

Note that this theorem does not depend on the dimension of the space.

10.18 **Orthogonality:** We define a vector ξ to be *orthogonal* to a vector η in a unitary space if $(\xi,\eta) = 0$. It is clear that if $(\xi,\eta) = 0$, then $(\eta,\xi) = \overline{(\xi,\eta)} = 0$ and if ξ is orthogonal to η, then η is also orthogonal to ξ.

Any set of orthogonal non-zero vectors in a unitary space is linearly independent. This is proved as in 6.13. If a set consists of vectors of norm one for which every two elements are orthogonal, we call that set orthonormal. If this set also generates the space, then it is called an orthonormal base.

10.19 **Theorem:** Let $\{\alpha_1, \ldots, \alpha_n\}$ be a set of n orthonormal vectors in a unitary space V. Let $\xi \in V$. Then

$$|(\xi,\alpha_1)|^2 + \ldots + |(\xi,\alpha_n)|^2 \leq |\xi|^2 \qquad \text{(Bessel's inequality).}$$

Also $\eta = \xi - [(\xi,\alpha_1)\,\alpha_1 + \ldots + (\xi,\alpha_n)\,\alpha_n]$ is orthogonal to α_i, $i = 1, \ldots, n$.

Proof: We have

$$|\eta|^2 = (\xi - [(\xi,\alpha_1)\alpha_1 + \ldots + (\xi,\alpha_n)\alpha_n\ ,\ \xi - [(\xi,\alpha_1)\alpha_1 + \ldots + (\xi,\alpha_n)\alpha_n])$$

$$= |\xi|^2 - [\overline{(\xi,\alpha_1)}(\xi,\alpha_1) + \ldots + \overline{(\xi,\alpha_n)}(\xi,\alpha_n) - [(\xi,\alpha_1)\overline{(\xi,\alpha_1)} + \ldots + (\xi,\alpha_n)\overline{(\xi,\alpha_n)}]$$

$$+ [(\xi,\alpha_1)\overline{(\xi,\alpha_1)} + \ldots + (\xi,\alpha_n)\overline{(\xi,\alpha_n)}] \geq 0$$

Therefore

$$|\xi|^2 - [|(\xi,\alpha_1)|^2 + \ldots + |(\xi,\alpha_n)|^2] \geq 0, \quad \text{or}$$

$$|\xi|^2 \geq |(\xi,\alpha_1)|^2 + \ldots + |(\xi,\alpha_n)|^2\ .$$

On the other hand

$$(\eta,\alpha_i) = (\xi,\alpha_i) - ((\xi,\alpha_1)\alpha_1 + \ldots + (\xi,\alpha_n)\alpha_n,\ \alpha_i)$$

$$= (\xi,\alpha_i) - (\xi,\alpha_i)(\alpha_i,\alpha_i) = 0,\ i = 1, \ldots, n\ .$$

An orthonormal set is called *complete* when the only vector orthogonal to all of its elements is the zero vector.

10.20 **Theorem:** In any finite dimensional unitary space E_n, a vector $\xi \epsilon E_n$ can be written in terms of the elements of an orthonormal base $\{\alpha_1, \ldots, \alpha_n\}$ as follows:

$$(1) \qquad \xi = (\xi, \alpha_1)\alpha_1 + \ldots + (\xi, \alpha_n)\alpha_n , \quad \text{and}$$

$$(2) \qquad |\xi|^2 = |(\xi, \alpha_1)|^2 + \ldots + |(\xi, \alpha_n)|^2 .$$

Proof: Since $\{\alpha_1, \ldots, \alpha_n\}$ is a base we have

$$\xi = a_1 \alpha_1 + \ldots + a_n \alpha_n , \quad \text{and}$$

$$(\xi, \alpha_i) = a_1(\alpha_1, \alpha_i) + \ldots + a_i(\alpha_i, \alpha_i) + \ldots + a_n(\alpha_i, \alpha_n) = a_i .$$

Therefore $(\xi, \alpha_i) = a_i$, $i = 1, \ldots, n$, and this proves (1). Now

$$|\xi|^2 = (\xi, \xi) = ((\xi, \alpha_1)\alpha_1 + \ldots + (\xi, \alpha_n)\alpha_n, (\xi, \alpha_1)\alpha_1 + \ldots + (\xi, \alpha_n)\alpha_n)$$

$$= (\xi, \alpha_1)\overline{(\xi, \alpha_1)} + \ldots + (\xi, \alpha_n)\overline{(\xi, \alpha_n)}$$

$$= |(\xi, \alpha_1)|^2 + \ldots + |(\xi, \alpha_n)|^2 .$$

Given any set of linearly independent vectors in a unitary space we can orthonormalize them in such a way that the orthonormal set obtained generates the same subspace. The proof is the same as in 6.15.

10.21 **Orthogonal complement of a subspace:** Let S be a subspace of a unitary space E over the complex field. A subspace T of E is called the orthogonal complement of S if

I $\quad S + T = E$,

II $\quad$ for $\xi \epsilon S$ and $\eta \epsilon T$ we have

$$(\xi, \eta) = 0 .$$

For any subspaces S and T of E satisfying II, if $\xi \epsilon S \cap T$, i.e., $\xi \epsilon S$ and $\xi \epsilon T$, then $(\xi, \xi) = |\xi|^2 = 0$, so that $\xi = 0$. Thus $S \cap T = 0$. If S is any subspace of E_n, a finite dimensional unitary space, there is a subspace T of E_n which is the orthogonal complement of S. The proof is left to the reader.

Clearly also any vector ζ of E can be written in one and only one way in the form

$$\zeta = \xi + \eta, \; \xi \epsilon S, \; \eta \epsilon T .$$

For since $S + T = E$, every ζ can be written in this form. But if $\zeta = \xi + \eta = \xi' + \eta'$, $\xi, \xi' \epsilon S$, $\eta, \eta' \epsilon T$, then $\xi - \xi' = \eta' - \eta$. But $\xi - \xi' \epsilon S$, $\eta' - \eta \epsilon T$, so that, as above, $\xi - \xi' = \eta' - \eta = 0$, that is, $\xi = \xi'$, $\eta = \eta'$.

EXERCISES 10

1. Show that the set of positive integers is closed under addition and multiplication, while the set of negative integers is not closed under multiplication.
2. Show that the set of all rotations in a three dimensional Euclidean space is a group (not commutative) under multiplication.
3. Show that the set of rigid motions in the plane is a group under multiplication.
4. Show that the set of all integers is a commutative group under addition.
5. Show that the set of all rational numbers is a field.
6. Show that the set of all numbers of the form $a + b\sqrt{c}$ where a and b are rational numbers and c is a fixed positive integer, is a field under addition and multiplication.
7. Show that the set of all polynomials of degree four with real coefficients is a subspace of the space of polynomial of degree $n > 4$.

8. Let V be the vector space of real continuous functions f on $[0, 1]$. Determine which of the following are subspaces of V.

   i all functions f such that $2f(0) = f(1)$.
   ii all functions f for which $f(1) = 1 + f(0)$.
   iii all functions f such that $f(x) = f(1 - x)$ for all x.

9. Show that $\{1, x, x^2, \ldots, x^n\}$ is a base for the vector space of all polynomials of degree n, with real coefficients and domain $[0,1]$.

10.* If the inner product for the space in 9 is defined to be

$$(P_1, P_2) = \int_0^1 [P_1(x)\, P_2(x)]\, dx ,$$

where P_1 and P_2 are two polynomials of degree less than or equal to n, orthonormalize the base in 9.

11. Show that the functions $\sin x$, $\sin 2x, \ldots, \sin nx$ on $[0,\pi]$ are linearly independent.

12.* If the inner product of y_1 and y_2, two functions of *ex.* 11 is defined to be

$$(y_1, y_2) = \frac{2}{\pi} \int_0^{\pi} y_1\, y_2\, dx ,$$

show that the set $\{\sin x, = \sin 2x, \ldots \sin nx\}$ is orthonormal.

13. Show that if $|\xi + \eta| = |\xi| + |\eta|$, then ξ and η are linearly dependent. (ξ, η are two vectors of a unitary space).

14. Let $\xi_1, \ldots, \xi_n$ be n vectors in an n-dimensional vector space V_n over a field F. Show that a necessary and sufficient condition for $\xi_1, \ldots \xi_n$ to be linearly dependent is

$$\begin{vmatrix} x_{11} & \ldots & x_{1n} \\ \ldots & & \ldots \\ x_{n1} & \ldots & x_{nn} \end{vmatrix} = 0 ,$$

where $(x_{i1}, \ldots, x_{in})$ is the set of coordinates of ξ_i, $i = 1, \ldots, n$ with respect to a base in V_n.

15. Let S be the subspace of V, a three-dimensional space over the field of real numbers, consisting of all vectors of the form $(0, y, z)$ and T the subspace spanned by (1,1,0) and (2,0,3). Describe $S \cap T$ and $S + T$.

16. Show that the set of all solutions (x_1, x_2, x_3) of

$$\begin{cases} a_{11} x_1 + a_{12} x_2 + a_{13} x_3 = 0 \\ a_{21} x_1 + a_{22} x_2 + a_{23} x_3 = 0 , \end{cases}$$

where a's and x's are all real, is a subspace of the three-dimensional space of all ordered triples (x_1, x_2, x_3) of real numbers.

17. Let S and T be two subspaces of a vector space V_n of dimension n over a field F, such that $\dim S = s$ and $\dim T = t$. Find the greatest possible dimension of $S + T$ and the least possible dimension of $S \cap T$.

11. LINEAR TRANSFORMATIONS IN GENERAL VECTOR SPACES

11.1 **Definitions:** Let V be a vector space over a field $\mathfrak{F}$. Then a *linear transformation* **A** on V is defined, as before, as a way of corresponding to each vector $\xi \epsilon$ V, a vector $\xi_1 = \mathbf{A}\,\xi \,\epsilon$ V, such that for any $\xi, \eta \,\epsilon$ V and $a, b \,\epsilon \mathfrak{F}$.

$$(1) \qquad \mathbf{A}(a\,\xi + b\,\eta) = a\mathbf{A}\,\xi + b\mathbf{A}\eta\ .$$

Again if **A** and **B** are two transformations, we may define the transformation $\mathbf{C} = \mathbf{A} + \mathbf{B}$ and $\mathbf{D} = \mathbf{AB}$, the *sum* and *product* of **A** and **B**, by the relations $\mathbf{C}\,\xi = \mathbf{A}\,\xi + \mathbf{B}\,\xi$, $\mathbf{D}\,\xi = (\mathbf{AB})\,\xi = \mathbf{B}(\mathbf{A}\,\xi)$, for all $\xi \,\epsilon$ V. The reader may verify that $\mathbf{A} + \mathbf{B}$ and **AB** are linear transformations satisfying (1), and it can be shown exactly as in 2.3 that for any transformations **A**, **B**, and **C** we have

$$(\mathbf{AB})\,\mathbf{C} = \mathbf{A}(\mathbf{BC})\ , \qquad \mathbf{A}(\mathbf{B} + \mathbf{C}) = \mathbf{AB} + \mathbf{AB}\ ;$$

that is, the multiplication is associative, and is distributive over addition. As before, we use the symbols **O** and **I** for the transformations defined by

$$\mathbf{O}\,\xi = 0\ , \qquad \mathbf{I}\,\xi = \xi$$

for all $\xi \,\epsilon$ V. These are called respectively the *zero* and *identity* (or unit) transformations.

11.2 **Space of linear transformations:** We observe from the above that if we consider the set of all linear transformations on a vector space V, it is, in particular, possible to add any two and get another in the set. Further, the transformation **O** has the property that $\mathbf{A} + \mathbf{O} = \mathbf{O} + \mathbf{A} = \mathbf{A}$ for every **A**. Let us define the transformation $-\mathbf{A}$ by the relation $(-\mathbf{A})\ \xi = -\ (\mathbf{A}\ \xi)$, and the transformation $k\mathbf{A}$, for $k \,\epsilon \mathfrak{F}$, by the relation $(k\mathbf{A})\ \xi = k(\mathbf{A}\ \xi)$ for all $\xi \,\epsilon$ V. It is clear then that this set of transformations forms an additive Abelian group, on which the field $\mathfrak{F}$ operates in such a way that

$$k(\mathbf{A} + \mathbf{B}) = k\mathbf{A} + k\mathbf{B}$$

for k in $\mathfrak{F}$, **A** and **B** any linear transformations. Thus the set of linear transformations on a vector space V over a field $\mathfrak{F}$ is itself a vector space over $\mathfrak{F}$.

11.3 **Algebra of linear transformations.** It is noted that a vector space (over a field $\mathfrak{F}$) in which a distributive (associative) multiplication is defined, is called a *linear* (*associative*) *algebra over* $\mathfrak{F}$. If there is an element **I** of this algebra with the property that $\mathbf{AI} = \mathbf{IA} = \mathbf{A}$ for every **A**, as is true for the set of linear transformations, this element **I** is called the (multiplicative) *unit* of the algebra.

We observe further that due to the associativity of the multiplication of transformations, we may define a transformation $\mathbf{A}^n$, where n is a positive integer, as the result of n successive applications of the transformation **A**; that is $\mathbf{A}^1 = \mathbf{A}$, $\mathbf{A}^n = \mathbf{A}^{n-1}\mathbf{A}$, $n>1$, and it will be true that for $a, b \,\epsilon \mathfrak{F}$, any integers m and n,

$$(2) \qquad (a \cdot \mathbf{A}^m) \cdot (b\mathbf{A}^n) = ab\mathbf{A}^{m+n}\ .$$

Then if we consider any polynomial function of a symbol x,

$$p(x) = a_o + a_1 x + \ldots + a_n x^n\ ,$$

where $a_o, a_1, \ldots, a_n \,\epsilon \mathfrak{F}$, we may define a linear transformation $p(\mathbf{A})$ by

$$p(\mathbf{A}) = a_o\mathbf{I} + a_1\mathbf{A} + \ldots + a_n\mathbf{A}^n\ .$$

From (2) it follows that the ordinary rules for multiplication of polynomials apply as well to the multiplication of polynomials in a transformation **A**. Since, however, the multiplication of transformations is not commutative, we will not attempt to define polynomials in several transformations, as the analogy to polynomials in two or more variables does not hold.

11.4 **Finite-dimensional vector spaces:** If a vector space V is of finite dimension n over a field $\mathcal{F}$, let $\{\alpha_1, \ldots, \alpha_n\}$ be a base for V. For a transformation **A**, let

$$\mathbf{A}\alpha_1 = a_{11}\alpha_1 + a_{12}\alpha_2 + \ldots + a_{1n}\alpha_n$$

$$\mathbf{A}\alpha_2 = a_{21}\alpha_1 + a_{22}\alpha_2 + \ldots + a_{2n}\alpha_n$$

$$\ldots$$

$$\mathbf{A}\alpha_n = a_{n1}\alpha_1 + a_{n2}\alpha_2 + \ldots + a_{nn}\alpha_n ,$$

where each $a_{ij} \in \mathcal{F}$. We may write these coefficients, called the components of $\mathbf{A}\alpha_i$ with respect to the base $\{\alpha_1, \ldots, \alpha_n\}$, in matrix form, thus obtaining the matrix of the transformation **A** with respect to the given base as in 7.2. When no confusion can arise, we will use the symbol **A** for the matrix as well as the transformation. Again, with respect to any base, the zero transformation has a matrix with every $a_{ij} = 0$, while the identity transformation has $a_{ii} = 1$, all other terms of the matrix being zero. It may be verified, as in 7.3, that addition and multiplication of transformations will correspond exactly to the addition and multiplication of their matrices. It is to be emphasized that the correspondence between a transformation **A** and its matrix depends on the choice of a particular base for V. However, it is clear that the set of all n-by-n matrices with elements in a field $\mathcal{F}$ forms a vector space over $\mathcal{F}$, and, in fact, a linear associative algebra over $\mathcal{F}$. The zero matrix is the zero element of the space, while if $\mathbf{A} = (a_{ij})$, the element $k\mathbf{A}$ is the matrix (b_{ij}) for which $b_{ij} = ka_{ij}$.

Let us consider now the n^2 matrices $\mathbf{E}_{ij}$ defined as follows. In the matrix $\mathbf{E}_{ij}$, let $a_{ij} = 1$, while all other terms are zero. Then for any matrix $\mathbf{A} = (a_{ij})$,

$$\mathbf{A} = a_{11}\mathbf{E}_{11} + a_{12}\mathbf{E}_{12} + \ldots + a_{1n}\mathbf{E}_{1n} + a_{21}\mathbf{E}_{21} + \ldots + a_{nn}\mathbf{E}_{nn} .$$

Furthermore, if for a set of n^2 elements $a_{ij} \in \mathcal{F}$

$$\mathbf{A} = a_{11}\mathbf{E}_{11} + \ldots + a_{1n}\mathbf{E}_{1n} + \ldots + a_{n1}\mathbf{E}_{n1} + \ldots + a_{nn}\mathbf{E}_{nn} = (\mathbf{0}),$$

then every a_{ij} must be zero. Thus the matrices $\mathbf{E}_{ij}$ form a base for the space of n-by-n matrices with elements in $\mathcal{F}$, so that this space is of dimension n^2 over $\mathcal{F}$.

In the algebra of matrices over $\mathcal{F}$, a special role is played by matrices corresponding to transformations of the form $a\mathbf{I}$ for $a \in \mathcal{F}$. The matrix of such a transformation is

$$[a] = [a\mathbf{I}] = \begin{pmatrix} a & & & 0 \\ & a & & \\ & & \ddots & \\ 0 & & & a \end{pmatrix},$$

and is called a scalar matrix, since for any matrix **A** it is true that $a\mathbf{A} = [a] \cdot \mathbf{A} = \mathbf{A} \cdot [a]$, so that these matrices operate in the algebra exactly as the scalar multipliers do in the vector space of n-by-n square matrices. Furthermore a polynomial $p(\mathbf{A}) = a_0\mathbf{I} + a_1\mathbf{A} + \ldots + a_n\mathbf{A}^n$ can be written as

$$p(\mathbf{A}) = [a_0] + [a_1]\,\mathbf{A} + \ldots + [a_n]\,\mathbf{A}^n .$$

11.5 **Rectangular matrices:** The definitions of rectangular matrices and their addition and multiplication, given in 2.8 remain valid for matrices over any field $\mathcal{F}$. The set of m-by-n matrices with elements in $\mathcal{F}$ is thus a vector space of dimension mn over $\mathcal{F}$, but not an algebra unless $m = n$, since the multiplication will not otherwise be defined. The components of the elements of a vector space of dimension n with respect to a fixed base may again be written as either 1-by-n or n-by-1 matrices.

11.6 **Rank and range of a transformation:** Let $\{\alpha_1, \alpha_2, \ldots, \alpha_n\}$ be a base for a vector space V over a field $\mathcal{F}$, and let **A** be a linear transformation on V. Let r be the rank [see 7.6] of the matrix of the transformation $\mathbf{A}$ with respect to the base $\{\alpha_1, \alpha_2, \ldots, \alpha_n\}$. We observe that the set of elements $\xi_1 \epsilon V$ which can be written in the form $\xi_1 = \mathbf{A}\xi$ is a subspace W of V, since $k\mathbf{A}\xi = \mathbf{A}(k\xi)$, and $\mathbf{A}\xi + \mathbf{A}\eta = \mathbf{A}(\xi+\eta)$. It can be shown that this subspace, called the *range* of $\mathbf{A}$, is of dimension r. For if the matrix of $\mathbf{A}$ is of rank r, then there are exactly r linearly independent rows in the matrix. Since these are the components of the vectors $\mathbf{A}\alpha_1, \ldots, \mathbf{A}\alpha_n$ with respect to the base $\{\alpha_1, \ldots, \alpha_n\}$, as in 10.10 there are exactly r linearly independent vectors in the set $\{\mathbf{A}\alpha_1, \ldots, \mathbf{A}\alpha_n\}$. But every element of W is a linear combination of $\mathbf{A}\alpha_1, \ldots, \mathbf{A}\alpha_n$, so that the dimension of W is r. The number $r = r(\mathbf{A})$ is called the *rank* of the transformation $\mathbf{A}$ since it is independent of the choice of base.

11.7 **Null space and nullity:** Let N be the set of elements $\xi \epsilon V$ such that $\mathbf{A}\xi = 0$. Clearly N is also a subspace of V, which may consist of only the zero-element of V. Then N is called the *null space* of the transformation $\mathbf{A}$, and its dimension $s = s(\mathbf{A})$ is called the *nullity* of $\mathbf{A}$. We prove that for any transformation $\mathbf{A}$ on V, $r(\mathbf{A}) + s(\mathbf{A}) = n$, that is the rank of $\mathbf{A}$ added to the nullity of $\mathbf{A}$ is equal to the dimension of the space V.

To prove this, let $\{\alpha_1, \ldots, \alpha_s\}$ be a base for N. Then by 10.13 there is a base $\{\alpha_1, \ldots, \alpha_s, \alpha_{s+1}, \ldots, \alpha_n\}$ for V which includes the base $\{\alpha_1, \ldots, \alpha_s\}$ of N. We show that $\mathbf{A}\alpha_{s+1}, \ldots \mathbf{A}\alpha_n$ are linearly independent and generate W. First if some $a_{s+1}\mathbf{A}\alpha_{s+1} + \ldots + a_n\mathbf{A}\alpha_n = 0$, $a_{s+1}, \ldots, a_n \epsilon \mathcal{F}$, then $\mathbf{A}(a_{s+1}\alpha_{s+1} + \ldots + a_n\alpha_n) = 0$, and $a_{s+1}\alpha_{s+1} + \ldots + a_n\alpha_n$, is in N. Then

$$a_{s+1}\alpha_{s+1} + \ldots + a_n\alpha_n = a_1\alpha_1 + \ldots + a_s\alpha_s, \text{ or } a_1\alpha_1 + \ldots + a_s\alpha_s - a_{s+1}\alpha_{s+1} - \ldots - a_n\alpha_n = 0 .$$

But since $\{\alpha_1, \ldots, \alpha_n\}$ is a base for V, this implies $a_i = 0$, $i = 1, 2, \ldots, n$, and in particular $a_{s+1} = \ldots = a_n = 0$. Further, if $\xi_1 \epsilon W$, then $\xi_1 = \mathbf{A}\xi = \mathbf{A}(a_1\alpha_1 + \ldots + a_s\alpha_s + a_{s+1}\alpha_{s+1} + \ldots + a_n\alpha_n) = a_1\mathbf{A}\alpha_1 + \ldots + a_s\mathbf{A}\alpha_s + a_{s+1}\mathbf{A}\alpha_{s+1} + \ldots + a_n\mathbf{A}\alpha_n = a_{s+1}\mathbf{A}\alpha_{s+1} + \ldots + a_n\mathbf{A}\alpha_n$. Since these n-s linearly independent vectors generate W, the dimension r of W is equal to n-s, that is,

$$r(\mathbf{A}) = n - s(\mathbf{A}) .$$

11.8 **Transform of a vector:** Let $\{\alpha_1, \ldots, \alpha_n\}$ be a base for a vector space V over a field $\mathcal{F}$. Let $\xi = x_1\alpha_1 + \ldots + x_n\alpha_n$ be an element of V, so that $(x_1 \ldots x_n)$ is the row matrix of the components of ξ with respect to $\{\alpha_1, \ldots, \alpha_n\}$. Then if $\xi_1 = \mathbf{A}\xi$ has components $(X_1, \ldots, X_n)$ in this base, we have as in 2.10,

$$(1) \qquad (X_1 \ldots X_n) = (x_1 \ldots x_n)\mathbf{A} ,$$

and

$$(2) \qquad \begin{pmatrix} X_1 \\ \cdot \\ \cdot \\ \cdot \\ X_n \end{pmatrix} = \mathbf{A}' \begin{pmatrix} x_1 \\ \cdot \\ \cdot \\ \cdot \\ x_n \end{pmatrix} ,$$

where by $\mathbf{A}'$ we mean the transpose of the matrix $\mathbf{A}$.

11.9 **Inverse of a transformation:** For a transformation $\mathbf{A}$ on a vector space V of dimension n over $\mathcal{F}$, if there is a transformation $\mathbf{B}$ such that $\mathbf{AB} = \mathbf{I}$, this transformation $\mathbf{B}$ is called an *inverse* of $\mathbf{A}$. Without reference to matrices, let us determine conditions that such an inverse exist. First, for any $\xi \epsilon V$, $(\mathbf{AB})\xi = \mathbf{B}(\mathbf{A}\xi) = \xi$, so that if $\mathbf{A}\xi = \xi_1$, $\mathbf{B}\xi_1 = \xi$. Hence if $\mathbf{A}\xi = \mathbf{A}\eta$, it must be true that $\xi = \eta$. This condition is obviously necessary for the existence of an inverse, and we will prove that it is also sufficient. We observe that if it is true that for every $\xi_1 \epsilon V$ there is a $\xi \epsilon V$ such that $\mathbf{A}\xi = \xi_1$, then the transformation $\mathbf{B}$ defined by $\mathbf{B}\xi_1 = \xi$ would have the property that $\mathbf{AB} = \mathbf{I}$. Let a base $\{\alpha_1, \ldots, \alpha_n\}$ be chosen in V. Since $\mathbf{A}\xi = \mathbf{A}\eta$ implies $\xi = \eta$, then $\mathbf{A}\xi = 0$ implies $\xi = 0$. Clearly, therefore ,$\mathbf{A}\alpha_1, \ldots, \mathbf{A}\alpha_n$ are linearly independent, since if $k_1\mathbf{A}\alpha_1 + \ldots + k_n\mathbf{A}\alpha_n = \mathbf{A}(k_1\alpha_1 + \ldots + k_n\alpha_n) = 0$, then $k_1\alpha_1 + \ldots + k_n\alpha_n = 0$, and

$k_i = 0$, $i = 1, \ldots, n$. These n linearly independent elements of V therefore form a base for V, and every $\xi_1 \epsilon$ V is a linear combination of them, that is, $\xi_1 = x_1 \mathbf{A}\alpha_1 + \ldots + x_n \mathbf{A}\alpha_n = \mathbf{A}(x_1\alpha_1 + \ldots + x_n\alpha_n) = \mathbf{A}\xi$, and the transformation $\mathbf{B}$ defined above exists. Clearly also $\mathbf{BA} = \mathbf{I}$, and we write $\mathbf{B} = \mathbf{A}^{-1}$.

The reader may show that if for two transformations $\mathbf{A}$ and $\mathbf{B}$, $\mathbf{A}^{-1}$ and $\mathbf{B}^{-1}$ exist then $(\mathbf{AB})^{-1}$ exists, and $(\mathbf{AB})^{-1} = \mathbf{B}^{-1}\mathbf{A}^{-1}$.

Now let us note that in the matrix of a transformation $\mathbf{A}$, since the rows of the matrix are the components of the vectors $\mathbf{A}\alpha_i$ with respect to the base $\{\alpha_1, \ldots, \alpha_n\}$, the number of vectors $\mathbf{A}\alpha_i$ which are linearly independent is equal to the rank of $\mathbf{A}$ or of its matrix, as in 7.6. Thus an inverse exists if and and only if the rank of the matrix is equal to the dimension n of V.

A transformation which has an inverse is called *non-singular*, as is a matrix of such a transformation. Otherwise it is called *singular*.

11.10 **Change of base:** Let $\{\alpha_1, \ldots, \alpha_n\}$ and $\{\beta_1, \ldots, \beta_n\}$ be two bases for V over $\mathcal{F}$. Then if

$$\beta_1 = d_{11}\,\alpha_1 + \ldots + d_{1n}\,\alpha_n\,,$$

$$\ldots$$

$$\beta_n = d_{n1}\,\alpha_1 + \ldots + d_{nn}\,\alpha_n\,,$$

We denote by $\mathbf{D}$ the matrix (d_{ij}), called the matrix of the change of base. Clearly $\mathbf{D}$ is non-singular and $\mathbf{D}^{-1}$ exists. If for $\xi \epsilon$ V, we have $\xi = x_1\alpha_1 + \ldots + x_n\alpha_n = X_1\,\beta_1 + \ldots + X_n\,\beta_n$, then as in 4.10,

$$(X_1 \ldots X_n) = (x_1 \ldots x_n)\,\mathbf{D}, \quad \text{and}$$

$$\begin{pmatrix} X_1 \\ \cdot \\ \cdot \\ \cdot \\ X_n \end{pmatrix} = \mathbf{D}' \begin{pmatrix} x_1 \\ \cdot \\ \cdot \\ \cdot \\ x_n \end{pmatrix}.$$

If $\mathbf{A}$ is a linear transformation whose matrix with respect to $\{\alpha_1, \ldots, \alpha_n\}$ is (a_{ij}), while its matrix with respect to $\{\beta_1, \ldots, \beta_n\}$ is (b_{ij}), then as in 4.10, $(b_{ij}) = \mathbf{D}(a_{ij})\,\mathbf{D}^{-1}$.

11.11 **Characteristic equation of a transformation:** Let $\mathbf{A}$ be a transformation on a vector space V of dimension n over a field $\mathcal{F}$. As in 5.1 and 7.12, if we wish to find non-zero elements ξ of V for which $\mathbf{A}\,\xi = k\,\xi$ for some $k \in \mathcal{F}$, we know that k must be a characteristic (proper) value of $\mathbf{A}$; that is, a solution in $\mathcal{F}$ of the characteristic equation of $\mathbf{A}$. If we consider the matrix of $\mathbf{A}$ with respect to some base, the characteristic equation may be written as

$$(1) \qquad |\mathbf{A} - m\mathbf{I}| = 0\,,$$

and in view of the results in 11.10, this equation is independent of the choice of base. It is a polynomial equation with coefficients in $\mathcal{F}$, of the form $p(m) = 0$. We call to the attention of the reader the fact that such an equation need not have any solutions in $\mathcal{F}$. For example, if $\mathcal{F}$ is the field of real numbers, V is the Euclidean two-dimensional space, and $\mathbf{A}$ is a rotation through an angle $\theta \neq n\pi$, the characteristic values of $\mathbf{A}$ are not in $\mathcal{F}$. Of course, if $\mathcal{F}$ is the complex field, the characteristic values are always in $\mathcal{F}$.

As before, a vector $\xi \epsilon$ V for which $\mathbf{A}\,\xi = k\,\xi$ is called a characteristic vector of A, corresponding to the proper value k.

11.12 **Cayley-Hamilton Theorem:** Let $p(m) = 0$ be the characteristic equation of a matrix $\mathbf{A}$. Then $p(\mathbf{A}) = 0$. That is, a matrix $\mathbf{A}$ satisfies its characteristic equation.

Proof: If $\mathbf{A} = (a_{ij})$, let us define C($\mathbf{A}$) to be the matrix (A_{ji}), where A_{ij} is the cofactor of a_{ij} in the determinant of $\mathbf{A}$, that is, C($\mathbf{A}$) is the transpose of the matrix of cofactors of $\mathbf{A}$. Then for any matrix $\mathbf{B}$,

$$\mathbf{B} \cdot C(\mathbf{B}) = \begin{pmatrix} |\mathbf{B}| & & & 0 \\ & |\mathbf{B}| & & \\ & & \ddots & \\ 0 & & & |\mathbf{B}| \end{pmatrix} = |\mathbf{B}| \cdot \mathbf{I} .$$

where $|\mathbf{B}|$ is the determinant of $\mathbf{B}$. For, since the columns of $C(\mathbf{B})$ are cofactors of the rows of $\mathbf{B}$, by 3.1 and 3.2 (5), the product matrix will have $|\mathbf{B}|$ along its main diagonal, and zero elsewhere.

Considering now the matrix $(\mathbf{A} - m\mathbf{I})$, we may form the matrix $C(\mathbf{A}-m\mathbf{I})$, the terms of which are polynomials in m of degree at most $n-1$. Thus, as in 11.3, $C(\mathbf{A}-m\mathbf{I})$ can be written as a polynomial in m with matrix coefficients, and

$$(\mathbf{A}-m\mathbf{I}) \cdot C(\mathbf{A}-m\mathbf{I}) = |\mathbf{A}-m\mathbf{I}| \cdot \mathbf{I} = p(m) \cdot \mathbf{I} .$$

This matrix equation, asserting the identity of two polynomials in the indeterminate m with matrix coefficients, will hold in particular if for m we substitute the matrix $\mathbf{A}$, whence

$$p(\mathbf{A}) \cdot \mathbf{I} = p(\mathbf{A}) = \mathbf{O},$$

and the theorem is established.

It can be shown that there is a polynomial $\phi(m)$ of minimum degree for which $\phi(\mathbf{A}) = \mathbf{O}$, called the *minimum function* of $\mathbf{A}$, which is unique except for a constant multiplier. Further, $\phi(m)$ is a divisor of $p(m)$ and of any other polynomial $\psi(m)$ for which $\psi(\mathbf{A}) = \mathbf{O}$. The determination and properties of the function $\phi(m)$ are discussed in detail in many standard works on vectors, matrices, and linear algebra, and will not be covered here.

We mention that a matrix $\mathbf{A}$ whose minimum function is of the form m^t, so that $\mathbf{A}^t = 0$, $\mathbf{A}^{t-1} \neq \mathbf{O}$, is called *nilpotent* of order t, as is a transformation whose matrix has this property.

11.13 **Unitary spaces and special transformations:** We consider now finite dimensional spaces over the complex field. Let E_n, then, be a space of dimension n over the complex field, in which an inner product (ξ, η) for $\xi, \eta \in E_n$ is defined, so that E_n is then a unitary space.

For a transformation $\mathbf{A}$ on E_n, let us define another transformation $\mathbf{A}^*$, called the *adjoint* of $\mathbf{A}$, by the relation

$$(\mathbf{A}\xi, \eta) = (\xi, \mathbf{A}^*\eta)$$

for all $\xi, \eta \in E_n$. As in 4.6 and 7.9,

$$(\mathbf{A}+\mathbf{B})^* = \mathbf{A}^* + \mathbf{B}^*, \qquad (\mathbf{AB})^* = \mathbf{B}^*\mathbf{A}^* .$$

If $\mathbf{A}$ is non-singular, so is $\mathbf{A}^*$, and

$$(\mathbf{A}^*)^{-1} = (\mathbf{A}^{-1})^* .$$

A transformation is called *normal* if $\mathbf{AA}^* = \mathbf{A}^*\mathbf{A}$. It is called *Hermitian* if $\mathbf{A} = \mathbf{A}^*$.

Again, a transformation $\mathbf{A}$ for which $(\mathbf{A}\xi, \mathbf{A}\eta) = (\xi, \eta)$, for every $\xi, \eta \in E_n$, is called *unitary*. For such a transformation clearly $(\mathbf{A}\xi, \mathbf{A}\eta) = (\xi, \mathbf{A}^*(\mathbf{A}\eta)) = (\xi, (\mathbf{AA}^*)\eta) = (\xi, \eta)$ so that

$$\mathbf{AA}^* = \mathbf{I}, \quad \text{or} \quad \mathbf{A}^* = \mathbf{A}^{-1} .$$

As in 4.9, the rows or columns of the matrix of a unitary transformation with respect to an orthonormal base themselves form an orthonormal set of vectors. The matrix is then also called unitary. As before, the matrix of a change of base from one orthonormal base to another is unitary.

Others of the theorems on Hermitian and unitary transformations which were stated earlier are true in general. For example, the characteristic values of a Hermitian transformation are real. Also, characteristic vectors corresponding to distinct characteristic values are orthogonal. Finally, there exists a base with respect to which the matrix of a Hermitian transformation has diagonal form. The proofs are exactly as in 7.14 and 7.15.

We observe also that the characteristic values of a unitary transformation all have absolute value 1. For if for some $\xi \in E_n$, and some complex number k, $\mathbf{A}\,\xi = k\,\xi$, and $\mathbf{A}$ is unitary, then

$$|\xi|^2 = (\xi, \xi) = (\mathbf{A}\xi, \mathbf{A}\xi) = (k\,\xi, k\,\xi) = k\bar{k}(\xi, \xi) = |k|^2\, |\xi|^2 ,$$

so that

$$|k| = 1 .$$

A Hermitian transformation $\mathbf{A}$ is called *non-negative* [*positive*] if for all $\xi \in E_n$ we have

$$(\mathbf{A}\xi, \xi) \geq 0 \quad [(\mathbf{A}\xi, \xi) > 0] .$$

It is left to the reader to show that the proper values of a non-negative [positive] transformation are non-negative [positive].

11.14 **Complementary subspaces:** In 10.21, we defined the orthogonal complement of a subspace S of a unitary space E. We wish now to generalize this notion for any vector space V, in which there may be no inner product, hence no concept of orthogonality. In general, then, two subspaces S and T of V will be called complementary if

I. $\quad S \cap T = 0$,

II. $\quad S + T = V$.

As in 10.21, it can then be shown that any vector $\zeta \in V$ can be written in one and only one way in the form $\zeta = \xi + \eta$, with $\xi \in S$, $\eta \in T$. As a footnote to 10.21, it may be stated that two subspaces S and T of a unitary space E are called orthogonal if $(\xi, \eta) = 0$ for every $\xi \in S$, $\eta \in T$.

11.15 **Projections:** For a transformation $\mathbf{P}$ on a vector space V, let us suppose there are two complementary subspaces S and T of V, such that if $\zeta \in V$, $\zeta = \xi + \eta$, $\xi \in S$, $\eta \in T$, then

$$\mathbf{P}\,\zeta = \xi .$$

Then the transformation $\mathbf{P}$ is called the projection of V onto the subspace S along T, and the reader may verify that $\mathbf{P}$ is a linear transformation. In particular $\mathbf{O}$ and $\mathbf{I}$ are projections.

We observe first that a projection $\mathbf{P}$ has the property that $\mathbf{P}^2 = \mathbf{P}$. For $\mathbf{P}^2\zeta = \mathbf{P}(\mathbf{P}\zeta) = \mathbf{P}\,\xi = \xi = \mathbf{P}\zeta$. A transformation $\mathbf{A}$ for which $\mathbf{A}^2 = \mathbf{A}$ is called *idempotent.*

The converse of the theorem above is also true; that is, if a transformation $\mathbf{P}$ is idempotent then it is a projection. The proof follows.

Let $\mathbf{P}^2 = \mathbf{P}$. Consider the set S of all vectors $\xi \in V$ such that $\xi = \mathbf{P}\,\zeta$ for $\zeta \in V$. Clearly, as in 11.6, the set S is a subspace of V. Likewise, the set T of vectors $\eta \in V$ which can be written in the form $\eta = \zeta - \mathbf{P}\,\zeta$ for $\zeta \in V$ is also a subspace of V. Further, $S + T = V$, since for any ζ,

$$\zeta = \mathbf{P}\,\zeta + (\zeta - \mathbf{P}\,\zeta) = \xi + \eta .$$

Also, for $\xi \in S$, $\mathbf{P}\,\zeta = \mathbf{P}(\mathbf{P}\zeta) = \mathbf{P}^2\,\zeta = \mathbf{P}\,\zeta = \xi$, while for $\eta \in T$, $\mathbf{P}\,\eta = \mathbf{P}(\zeta - \mathbf{P}\zeta) = \mathbf{P}\,\zeta - \mathbf{P}^2\,\zeta = 0$. Thus if $\alpha \in S \cap T$, since $\alpha \in S$, $\mathbf{P}\,\alpha = \alpha$, and since $\alpha \in T$, $\mathbf{P}\alpha = 0$, so that $\alpha = 0$. Then S and T are complementary subspaces of V, and $\mathbf{P}$ is the projection of V onto S along T. Note that S consists of all elements $\xi \in V$ for which $\mathbf{P}\,\xi = \xi$, and T consists of all $\eta \in V$ such that $\mathbf{P}\,\eta = 0$.

The transformation $\mathbf{Q}$ defined by $\mathbf{Q}\,\zeta = \eta$ is clearly also a projection, onto T along S, and the above argument shows, therefore, that a transformation $\mathbf{P}$ is a projection if and only if $\mathbf{Q} = \mathbf{I} - \mathbf{P}$ is a projection. This trivial fact is of some use in discussing the algebra of transformations as in 11.3, as it applies to projections.

11.16 **Algebra of projections:** If $\mathbf{P}_1$ and $\mathbf{P}_2$ are projections of a vector space V, onto S_1 along T_1, and onto S_2 along T_2 respectively, then

I. $\mathbf{P}_1 + \mathbf{P}_2$ is a projection if and only if $\mathbf{P}_1\mathbf{P}_2 = \mathbf{P}_2\mathbf{P}_1 = \mathbf{O}$. Then $S_1 \cap S_2 = 0$, and $\mathbf{P}_1 + \mathbf{P}_2 = \mathbf{P}$ is the projection onto $S = S_1 + S_2$ along $T = T_1 \cap T_2$.

II. If $\mathbf{P}_1\mathbf{P}_2 = \mathbf{P}_2\mathbf{P}_1 = \mathbf{P}$, then $\mathbf{P}$ is the projection onto $S = S_1 \cap S_2$ along $T = T_1 + T_2$.

To prove I, we observe that if $\mathbf{P}_1\mathbf{P}_2 = \mathbf{P}_2\mathbf{P}_1 = \mathbf{O}$, then

$$(\mathbf{P}_1 + \mathbf{P}_2)^2 = \mathbf{P}_1^2 + \mathbf{P}_2^2 + \mathbf{P}_1\mathbf{P}_2 + \mathbf{P}_2\mathbf{P}_1 = \mathbf{P}_1^2 + \mathbf{P}_2^2 = \mathbf{P}_1 + \mathbf{P}_2 .$$

Conversely, since in any case $(\mathbf{P}_1 + \mathbf{P}_2)^2 = \mathbf{P}_1^2 + \mathbf{P}_2^2 + \mathbf{P}_1\mathbf{P}_2 + \mathbf{P}_2\mathbf{P}_1$, and $\mathbf{P}_1^2 + \mathbf{P}_2^2 = \mathbf{P}_1 + \mathbf{P}_2$, if $(\mathbf{P}_1 + \mathbf{P}_2)^2 = \mathbf{P}_1 + \mathbf{P}_2$, we must have

(1) $\quad \mathbf{P}_1\mathbf{P}_2 + \mathbf{P}_2\mathbf{P}_1 = \mathbf{O}$.

Then $\mathbf{P}_1\mathbf{P}_1\mathbf{P}_2 + \mathbf{P}_1\mathbf{P}_2\mathbf{P}_1 = \mathbf{O} = \mathbf{P}_1\mathbf{P}_2\mathbf{P}_1 + \mathbf{P}_2\mathbf{P}_1\mathbf{P}_1$, so that

$$\mathbf{P}_1\mathbf{P}_2 = \mathbf{P}_1\mathbf{P}_1\mathbf{P}_2 = \mathbf{P}_2\mathbf{P}_1\mathbf{P}_1 = \mathbf{P}_2\mathbf{P}_1 ,$$

which, with (1), gives

$$\mathbf{P}_1\mathbf{P}_2 = \mathbf{P}_2\mathbf{P}_1 = \mathbf{O} .$$

We now observe first that if $\xi_1 \epsilon S_1$, then $\mathbf{P}_1\,\xi_1 = \xi_1$, so that $\mathbf{P}_2\,\xi_1 = \mathbf{P}_2(\mathbf{P}_1\,\xi_1) = (\mathbf{P}_1\mathbf{P}_2)\,\xi_1 = 0$, and similarly if $\xi_2 \epsilon S_2$, then $\mathbf{P}_1\,\xi_2 = 0$. Thus $S_1 \cap S_2 = 0$. Next, if $\xi \epsilon S = S_1 + S_2$, ξ can be written in the form $\xi = \xi_1 + \xi_2$, $\xi_1 \epsilon S_1$, $\xi_2 \epsilon S_2$. Then $\mathbf{P}\,\xi = \mathbf{P}_1\,\xi + \mathbf{P}_2\,\xi = \mathbf{P}_1\,\xi_1 + \mathbf{P}_2\,\xi_2 = \xi_1 + \xi_2 = \xi$. We observe that for any $\zeta \epsilon V$, $\mathbf{P}\,\zeta = \mathbf{P}_1\,\zeta + \mathbf{P}_2\,\zeta$, and $\mathbf{P}_1\,\zeta \epsilon S_1$, $\mathbf{P}_2\,\zeta \epsilon S_2$ so that $\mathbf{P}\,\zeta \epsilon S_1 + S_2$. In particular, if $\zeta = \mathbf{P}\,\zeta$, then $\zeta \epsilon S_1 + S_2$, and $S = S_1 + S_2$ consists of precisely those elements ζ of V for which $\mathbf{P}\,\zeta = \zeta$. Finally, if $\mathbf{P}\,\zeta = 0$, $\mathbf{P}_1\,\zeta = -\mathbf{P}_2\,\zeta \epsilon S_1 \cap S_2$, so that $\mathbf{P}_1\,\zeta = \mathbf{P}_2\,\zeta = 0$, and $\zeta \epsilon T = T_1 \cap T_2$, while clearly if $\zeta \epsilon T$, $\mathbf{P}\,\zeta = 0$. Then $\mathbf{P}_1 + \mathbf{P}_2$ is the projection onto S along T.

As for II, obviously if $\mathbf{P} = \mathbf{P}_1\mathbf{P}_2 = \mathbf{P}_2\mathbf{P}_1$, then $\mathbf{P}^2 = (\mathbf{P}_1\mathbf{P}_2)(\mathbf{P}_1\mathbf{P}_2) = \mathbf{P}_1(\mathbf{P}_2\mathbf{P}_1)\mathbf{P}_2 = \mathbf{P}_1(\mathbf{P}_1\mathbf{P}_2)\mathbf{P}_2 = \mathbf{P}_1^2\mathbf{P}_2^2 = \mathbf{P}_1\mathbf{P}_2 = \mathbf{P}$, so that $\mathbf{P}$ is a projection. The proof of the rest is similar to that for I, and is left as an exercise for the reader.

11.17 **Matrix of a projection:** If $\mathbf{P}$ is the projection of a finite-dimensional vector space V onto a subspace S along a complementary subspace T, it is possible to choose a base for V in which the matrix of $\mathbf{P}$ takes a very simple form. Let a base $\{\alpha_1, \ldots, \alpha_k\}$ be chosen for S, and a base $\{\beta_1, \ldots, \beta_{n-k}\}$ be chosen for T. Then $\mathbf{P}\,\alpha_i = \alpha_i$, $i = 1, \ldots, k$, and $\mathbf{P}\,\beta_j = 0$, $j = 1, \ldots, n-k$. But the set $\{\alpha_1, \ldots, \alpha_k, \beta_1, \ldots, \beta_{n-k}\}$ is a base for V. For since $V = S + T$, this set clearly generates V. On the other hand, if some $a_1\,\alpha_1 + \ldots + a_k\,\alpha_k + b_1\,\beta_1 + \ldots + b_{n-k}\,\beta_{n-k} = 0$, then since $S \cap T = 0$, $a_1\,\alpha_1 + \ldots + a_k\,\alpha_k = -b_1\,\beta_1 - \ldots - b_{n-k}\,\beta_{n-k} = 0$, so that $a_1 = \ldots = a_k = b_1 = \ldots = b_{n-k} = 0$. Then with respect to this base the matrix takes the form (a_{ij}), where $a_{ij} = 0$ if $i \neq j$, $a_{ii} = 1$ for $i = 1, \ldots, k$, $a_{ii} = 0$ for $i = k+1, \ldots, n$, that is

$$\mathbf{P} = \begin{pmatrix} [\mathbf{I}] & [\mathbf{O}] \\ [\mathbf{O}] & [\mathbf{O}] \end{pmatrix} .$$

11.18 **Perpendicular projection:** It was indicated in 10.21 that for any subspace S of a unitary space E over the complex field, it was possible to find a subspace T which was the orthogonal complement of S. In particular in this case S and T are complementary subspaces of E, and the projection of E onto S along T is called the *perpendicular projection* of E onto S. Further, since in this case the bases for S and T, as in 11.17, may be chosen as orthonormal, and the base thus constructed for E will then also be orthonormal, we find that with respect to this base the matrix of the transformation $\mathbf{P}$ is in diagonal form, where the terms on the main diagonal are all either 1 or 0. Clearly, therefore, a perpendicular projection is a non-negative Hermitian transformation.

11.19 **Decomposition of Hermitian transformations:** If E is a finite-dimensional unitary space over the complex field, and $\mathbf{A}$ is a Hermitian transformation on E, it was observed in 11.13 that the characteristic roots (proper values) of $\mathbf{A}$ are all real, and that vectors corresponding to distinct characteristic roots are orthogonal. Further if m is a characteristic root of $\mathbf{A}$, then the set S of vectors ξ of E for which $\mathbf{A}\,\xi = m\,\xi$ is a subspace of E with dimension equal to the multiplicity of the characteristic root m. It is obvious, furthermore, that the subspace generated by all characteristic vectors corresponding to the other characteristic roots is the orthogonal complement of S.

Now let $m_1, m_2, \ldots, m_k$ be the distinct characteristic roots of $\mathbf{A}$. Let S_i, $i = 1, \ldots, k$ be the subspace of vectors $\xi \epsilon$ E for which $\mathbf{A}\ \xi = m_i\ \xi$. Clearly for $i \neq j$, S_i and S_j are orthogonal subspaces of E. Further, if $\{\alpha_{i1}, \ldots, \alpha_{i,d_i}\}$ is an orthonormal base for S_i, where d_i is then the dimension of S_i, then the set $\{\alpha_{ij}\}$ consisting of all the α's is an orthonormal base for E. Finally let $\mathbf{P}_i$ be the perpendicular projection of E onto S_i. Then with respect to the base above, the matrix of A has the form

$$\begin{pmatrix} m_1 & & & & & & & & \\ & \ddots & & & & & & & \\ & & m_1 & & & & & 0 & \\ & & & m_2 & & & & & \\ & & & & \ddots & & & & \\ & & & & & m_2 & & & \\ & 0 & & & & & \ddots & & \\ & & & & & & & m_k & \\ & & & & & & & & \ddots \\ & & & & & & & & & m_k \end{pmatrix} = \begin{pmatrix} [m_1\mathbf{I}_{d_1}] & 0 & \cdots & 0 \\ 0 & [m_2\mathbf{I}_{d_2}] & & \cdot \\ \cdot & & \cdot & \cdot \\ \cdot & & \cdot & \cdot \\ \cdot & & \cdot & 0 \\ 0 & \cdots & 0 & [m_k\mathbf{I}_{d_k}] \end{pmatrix},$$

where $\mathbf{I}_{d_i}$ means the d_i–by–d_i identity matrix. On the other hand, relative to this base, the projection $\mathbf{P}_i$ has the matrix

$$\begin{pmatrix} [\mathbf{O}_{d_1}] & \cdots & & 0 \\ 0 & \ddots & & \vdots \\ \vdots & & [\mathbf{I}_{d_i}] & \vdots \\ \vdots & & \ddots & 0 \\ 0 & \cdots & & [\mathbf{O}_{d_k}] \end{pmatrix}$$

Thus the Hermitian transformation $\mathbf{A}$ can be written as a linear combination of projections,

$$\mathbf{A} = m_1\mathbf{P}_1 + m_2\mathbf{P}_2 + \ldots + m_k\mathbf{P}_k\ .$$

EXERCISES 11

1. Let $\{j_1, \ldots, j_n\}$ be a fixed permutation of $\{1, 2, \ldots, n\}$, the set of the first n positive integers. For any vector $\xi = (x_1, \ldots, x_n)$ of a complex n-dimensional space let $\mathbf{A}\ \xi = (x_{j_1}, \ldots, x_{j_n})$. Show that $\mathbf{A}$ is a linear transformation, and write its matrix.
2. Let $\xi \epsilon\ E_n$, an n-dimensional unitary space over the complex field, and let $\{\alpha_1, \ldots, \alpha_n\}$ be an orthonormal base for E_n. Let a transformation A be defined by $A\ \xi = a_1\,(\xi, \alpha_1)\,\alpha_1 + \ldots + a_k\,(\xi, \alpha_k)\,\alpha_k$, where $k < n$, and $a_1, \ldots, a_k$ is a fixed set of complex numbers. Show that $\mathbf{A}$ is a linear transformation, and write the matrix of $\mathbf{A}$.

3. Let ξ be as in 2., and $\mathbf{A}\,\xi = [a_{11}\,(\xi,\,\alpha_1) + \ldots + a_{1k}\,(\xi,\,\alpha_k)]\,\alpha_1 + \ldots + [a_{k1}\,(\xi,\,\alpha_1) + \ldots + a_{kk}\,(\xi,\,\alpha_k)]\,\alpha_k$, where $k < n$, and a_{ij}, $i, j = 1, \ldots, k$, are complex numbers. Show that $\mathbf{A}$ is a linear transformation, and write the matrix of $\mathbf{A}$. What is a necessary and sufficient condition on the numbers a_{ij} that $\mathbf{A}$ be a projection?
4. Find the rank, range, null space, and nullity of the transformations described in 2. and 3.
5. Show that a projection $\mathbf{P}$ is a perpendicular projection if and only if $\mathbf{P}^* = \mathbf{P}$, that is, $\mathbf{P}$ is Hermitian.
6. Show that in the complex Euclidean three-space, the transformation $\mathbf{A}$ with matrix

$$\begin{pmatrix} \frac{8}{9} & \frac{-\sqrt{3}-i}{9} & \frac{-1-i\sqrt{3}}{9} \\ \frac{-\sqrt{3}+i}{9} & \frac{5}{9} & \frac{-2\sqrt{3}-2i}{9} \\ \frac{-1+i\sqrt{3}}{9} & \frac{-2\sqrt{3}+2i}{9} & \frac{5}{9} \end{pmatrix}$$

is a projection, and find its range and null space. Change the matrix to diagonal form.

7. Show that as in 6. the matrix

$$\begin{pmatrix} \frac{45}{49} & -\frac{6}{49}\,e^{i\theta} & -\frac{12}{49}\,e^{i\phi} \\ -\frac{6}{49}\,e^{-i\theta} & \frac{40}{49} & -\frac{18}{49}\,e^{i(\phi-\theta)} \\ -\frac{12}{49}\,e^{-i\phi} & -\frac{18}{49}e^{-(\phi-\theta)} & \frac{13}{49} \end{pmatrix}$$

is the matrix of a projection, and find its range and null space.

8. If $\mathbf{A}$ is any linear transformation on a unitary space over the complex field, show that the characteristic values of $\mathbf{A}^*$ are the conjugates of the characteristic values of $\mathbf{A}$.
9. Prove that if $\mathbf{A}$ is normal, then ξ is a characteristic vector of $\mathbf{A}$ if and only if it is a characteristic vector of $\mathbf{A}^*$. Also if $\mathbf{A}\,\xi = \lambda\,\xi$, then $\mathbf{A}^*\,\xi = \bar{\lambda}\,\xi$.
10. Prove that if $\mathbf{A}$ is normal, then characteristic vectors corresponding to distinct characteristic roots are orthogonal.
11. Show that in a finite-dimensional unitary space E_n, there exists a base with respect to which a given normal transformation has a matrix in diagonal form.
12. Using the result of 11., show that a normal transformation can be written as a linear combination of projections onto a set of orthogonal subspaces $S_1, S_2, \ldots, S_k$, such that $S_1 + S_2 + \ldots + S_k = E_n$.

12. SINGULAR VALUES AND ESTIMATES OF PROPER VALUES OF MATRICES

12.1 **Proper values of a matrix:** In this chapter we consider linear transformations always on E_n, a unitary space of dimension n over the complex field. As in 11.11, a proper value of a linear transformation $\mathbf{A}$ is a scalar λ satisfying the equation

$$|\mathbf{A} - \lambda \mathbf{I}| = 0 .$$

In general λ is a complex number. For a Hermitian transformation λ is real, for non-negative (positive) transformations, λ is non-negative (positive), and for a unitary transformation, $|\lambda| = 1$ (see 11.13).

12.2 **Theorem:** (*Fischer's Minimax Principle*): Let $\mathbf{A}$ be a Hermitian transformation with proper values $\lambda_1 \geq \ldots \geq \lambda_n$. Then

$$(1) \qquad \lambda_1 = \sup_{|\xi|=1} (\mathbf{A}\xi, \xi) ,$$

$$(2) \qquad \lambda_n = \inf_{|\xi|=1} (\mathbf{A}\xi, \xi) ,$$

$$(3) \qquad \lambda_k = \sup_{\substack{M \\ \dim M = k}} \ \inf_{\substack{|\xi|=1 \\ \xi \in M}} (\mathbf{A}\xi, \xi) ,$$

where M is a subspace of E_n.

Proof: Let $\{\alpha_1, \ldots, \alpha_n\}$ be an orthonormal set of proper vectors of $\mathbf{A}$ corresponding to $\lambda_1, \ldots, \lambda_n$ respectively. Then for any $\xi \in E_n$ we have, as in 10.20,

$$\xi = (\xi, \alpha_1)\, \alpha_1 + \ldots + (\xi, \alpha_n)\, \alpha_n, \quad \text{and}$$

$$\mathbf{A}\xi = (\xi, \alpha_1)\, \mathbf{A}\alpha_1 + \ldots + (\xi, \alpha_n)\, \mathbf{A}\alpha_n \quad \text{or}$$

$$\mathbf{A}\xi = (\xi, \alpha_1)\, \lambda_1 \alpha_1 + \ldots + (\xi, \alpha_n)\, \lambda_n \alpha_n .$$

Now if $|\xi| = 1$, we have

$$(\mathbf{A}\xi, \xi) = |(\xi, \alpha_1)|^2 \lambda_1 (\alpha_1, \alpha_1) + \ldots + |(\xi, \alpha_n)|^2 \lambda_n (\alpha_n, \alpha_n)$$

$$= \lambda_1 |(\xi, \alpha_1)|^2 + \ldots + \lambda_n |(\xi, \alpha_n)|^2 \leq \lambda_1 [|(\xi, \alpha_1)|^2 + \ldots + |(\xi, \alpha_n)|^2]$$

$$= \lambda_1 |\xi|^2 = \lambda_1, \quad \text{so that} \quad \sup (\mathbf{A}\xi, \xi) \leq \lambda_1 .$$

But $(\mathbf{A}\alpha_1, \alpha_1) = \lambda_1$, which proves (1).

On the other hand, we see that for $|\xi| = 1$,

$$(\mathbf{A}\xi, \xi) = \lambda_1 |(\xi, \alpha_1)|^2 + \ldots + \lambda_n |(\xi, \alpha_n)|^2 \geq \lambda_n [|(\xi, \alpha_1)|^2 + \ldots + |(\xi, \alpha_n)|^2] = \lambda_n ,$$

and $(\mathbf{A}\alpha_n, \alpha_n) = \lambda_n$, which proves (2).

Now let M be the subspace of E_n generated by $\{\alpha_1, \ldots, \alpha_k\}$. Then for any $\xi \in M$ with $|\xi| = 1$ we have $\xi = (\xi, \alpha_1)\, \alpha_1 + \ldots + (\xi, \alpha_k)\, \alpha_k$. Then

$$(\mathbf{A}\xi, \xi) = |(\xi, \alpha_1)|^2 \lambda_1 + \ldots + |(\xi, \alpha_k)|^2 \lambda_k \geq \lambda_k [|(\xi, \alpha_1)|^2 + \ldots + |(\xi, \alpha_k)|^2] = \lambda_k |\xi|^2 = \lambda_k .$$

Thus for $\xi \in M$, $(\mathbf{A}\xi, \xi)$ is at least equal to λ_k.

On the other hand, now let M be any k-dimensional subspace of E_n. Let N be the subspace generated by $\{\alpha_k, \alpha_{k+1}, \ldots, \alpha_n\}$. Clearly dim M = k, and dim N = $n-k+1$. Also by 10.15, dim M + dim N = dim (M + N) + dim (M $\cap$ N). But since dim M + dim N $= k + (n - k + 1) = n + 1$, and dim (M + N) $\leq n$, we have dim (M $\cap$ N) ≥ 1. Therefore there exists a vector $\xi \epsilon$ M such that $\xi \epsilon$ N and $|\xi| = 1$. For this ξ,

$$\xi = (\xi,\alpha_k)\alpha_k + \ldots + (\xi,\alpha_n)\alpha_n, \quad \text{and}$$

$$(\mathbf{A}\xi,\xi) = \lambda_k|(\xi,\alpha_k)|^2 + \ldots + \lambda_n|(\xi,\alpha_n)|^2 \leq \lambda_k[|(\xi,\alpha_k)|^2 + \ldots + |(\xi,\alpha_n)|^2] = \lambda_k|\xi|^2 = \lambda_k .$$

This shows that

$$\inf_{\substack{|\xi|=1 \\ \xi \epsilon M}} (\mathbf{A}\xi,\xi) \leq \lambda_k ,$$

and therefore

$$\sup_{\substack{M \\ \dim M = k}} \; \inf_{\substack{|\xi|=1 \\ \xi \epsilon M}} (\mathbf{A}\xi,\xi) \leq \lambda_k .$$

Therefore

$$\lambda_k = \sup_{\substack{M \\ \dim M = k}} \; \inf_{\substack{\xi \epsilon M \\ |\xi|=1}} (\mathbf{A}\xi,\xi) .$$

12.3 **Cartesian decomposition of a linear transformation:** Linear transformations are analogous to complex numbers in many ways. Among the more important similarities are these. The operation which leads from a matrix to its adjoint corresponds to obtaining the conjugate of a complex number. Hermitian transformations play the part of real numbers. A unitary transformation is similar to a complex number of unit absolute value. Now any linear transformation $\mathbf{A}$ on E_n can be decomposed uniquely as $\mathbf{A} = \mathbf{B} + i\mathbf{C}$, where $i = \sqrt{-1}$, and $\mathbf{B}$ and $\mathbf{C}$ are Hermitian.

In fact,

$$\mathbf{B} = \frac{\mathbf{A} + \mathbf{A}^*}{2} \quad \text{and} \quad \mathbf{C} = \frac{\mathbf{A} - \mathbf{A}^*}{2i} ,$$

i.e.,

$$\mathbf{A} = \left(\frac{\mathbf{A} + \mathbf{A}^*}{2}\right) + i\left(\frac{\mathbf{A} - \mathbf{A}^*}{2i}\right) .$$

It is easily seen that

$$\left(\frac{\mathbf{A} + \mathbf{A}^*}{2}\right)^* = \frac{\mathbf{A} + \mathbf{A}^*}{2} , \quad \text{and}$$

$\lambda_1 \geq \cdots \geq \lambda_n$ real singular values of A

$$\left(\frac{\mathbf{A} - \mathbf{A}^*}{2i}\right)^* = \frac{\mathbf{A} - \mathbf{A}^*}{2i} ,$$

$\mu_1 \geq \cdots \geq \mu_n$ imaginary singular values of A

which means that $\mathbf{B}$ and $\mathbf{C}$ are Hermitian. It is left to the reader to prove the uniqueness of this decomposition.

The analogy between linear transformations and complex numbers becomes deeper when we consider the proper values of transformations or matrices. A Hermitian transformation has only real proper values, and the proper values of non-negative transformations are non-negative. The proper values of a unitary transformation have modulus one. Since more is known about the proper values of Hermitian transformations, we will consider here certain Hermitian transformations related to a given transformation $\mathbf{A}$, and we shall estimate the real and imaginary parts, and the absolute values of the proper values of a linear transformation in terms of the proper values of those Hermitian transformations.

12.4 **Singular values of a transformation:** The proper values of $\frac{\mathbf{A}+\mathbf{A}^*}{2}$ are called the *real singular values* of $\mathbf{A}$, and the proper values of $\frac{\mathbf{A}-\mathbf{A}^*}{2i}$ are called the *imaginary singular values* of $\mathbf{A}$.

Before we define the absolute singular values of $\mathbf{A}$, we show that $\mathbf{A}^*\mathbf{A}$ and $\mathbf{A}\mathbf{A}^*$ are non-negative and both have the same proper values.

It is clear that $(\mathbf{A}^*\mathbf{A})^* = \mathbf{A}^*\mathbf{A}$, and $(\mathbf{A}\mathbf{A}^*)^* = \mathbf{A}\mathbf{A}^*$. Now for any non-zero $\alpha \epsilon E_n$ we have $(\mathbf{A}\mathbf{A}^*\alpha,\alpha) = (\mathbf{A}^*[\mathbf{A}\alpha], \alpha) = (\mathbf{A}\alpha,\mathbf{A}\alpha) = |\mathbf{A}\alpha|^2 \geq 0$, and $(\mathbf{A}^*\mathbf{A}\alpha,\alpha) = (\mathbf{A}^*\alpha,\mathbf{A}^*\alpha) = |\mathbf{A}^*\alpha|^2 \geq 0$, i.e., $\mathbf{A}^*\mathbf{A}$ and $\mathbf{A}\mathbf{A}^*$ are non-negative transformations. Now suppose α is a proper vector of $\mathbf{A}\mathbf{A}^*$ corresponding to a proper value λ of $\mathbf{A}\mathbf{A}^*$. Then

$$\mathbf{A}\mathbf{A}^*\,\alpha = \lambda\,\alpha.$$

Operating with $\mathbf{A}$ on both sides of the above equality, we get

$$\mathbf{A}(\mathbf{A}\mathbf{A}^*\alpha) = \mathbf{A}\mathbf{A}^*\mathbf{A}\alpha = \mathbf{A}(\lambda\alpha) = \lambda\mathbf{A}\,\alpha.$$

Let $\mathbf{A}\alpha = \beta$. Then since $\mathbf{A}\mathbf{A}^*\mathbf{A}\alpha = \mathbf{A}^*\mathbf{A}(\mathbf{A}\alpha)$,

$$\mathbf{A}^*\mathbf{A}\beta = \lambda\beta.$$

Thus any proper value of $\mathbf{A}\mathbf{A}^*$ is a proper value of $\mathbf{A}^*\mathbf{A}$. Now let β be a proper vector of $\mathbf{A}^*\mathbf{A}$ corresponding to μ, a proper value of $\mathbf{A}^*\mathbf{A}$. Then

$$\mathbf{A}^*\mathbf{A}\,\beta = \mu\,\beta, \quad \text{and}$$

$$\mathbf{A}^*\mathbf{A}\mathbf{A}^*\,\beta = \mu\,A^*\beta,$$

and writing $A^*\beta = \alpha$ we get

$$AA^*\,\alpha = \mu\,\alpha,$$

i.e., any proper value of $\mathbf{A}^*\mathbf{A}$ is also a proper value of $\mathbf{A}\mathbf{A}^*$.

Now we are in a position to define the absolute singular values of $\mathbf{A}$. The non-negative square roots of the proper values of $\mathbf{A}\mathbf{A}^*$ are called the *absolute singular values* of $\mathbf{A}$.

12.5 **Theorem:** Let $\mathbf{A}$ be a Hermitian transformation on E_n with proper values $\lambda_1, \ldots, \lambda_n$. Let $\{\xi_1, \ldots, \xi_n\}$ be any orthonormal set of vectors in E_n. Then

$$(\mathbf{A}\xi_1, \xi_1) + \ldots + (\mathbf{A}\xi_n, \xi_n) = \lambda_1 + \ldots + \lambda_n.$$

Proof: Let $\{\alpha_1, \ldots, \alpha_n\}$ be an orthonormal set such that $\mathbf{A}\alpha_i = \lambda_i\,\alpha_i$, $i = 1, \ldots, n$. Then

$$\xi_1 = (\xi_1,\alpha_1)\,\alpha_1 + \ldots + (\xi_1,\alpha_n)\,\alpha_n,$$

$$\ldots \qquad \ldots$$

$$\xi_n = (\xi_n,\alpha_1)\,\alpha_1 + \ldots + (\xi_n,\alpha_n)\,\alpha_n.$$

Therefore

$$(\mathbf{A}\,\xi_1, \xi_1) = \lambda_1\,|(\xi_1, \alpha_1)|^2 + \ldots + \lambda_n\,|(\xi_1, \alpha_n)|^2$$

$$\ldots \qquad \ldots$$

$$(\mathbf{A}\,\xi_n, \xi_n) = \lambda_1\,|(\xi_n, \alpha_1)|^2 + \ldots + \lambda_n\,|(\xi_n, \alpha_n)|^2.$$

Thus

$$(\mathbf{A}\,\xi_1, \xi_1) + \ldots + (\mathbf{A}\xi_n, \xi_n) = \lambda_1[|(\xi_1, \alpha_1)^2 + \ldots + |(\xi_n, \alpha_1)|^2]$$
$$+ \ldots + \lambda_n[|(\xi_1, \alpha_n)|^2 + \ldots + |(\xi_n, \alpha_n)|^2].$$

But $|(\xi_1, \alpha_i)|^2 + \ldots + |(\xi_n, \alpha_i)|^2 = |\alpha_i|^2 = 1$, $i = 1, \ldots, n$, and

$$(\mathbf{A}\,\xi_1, \xi_1) + \ldots + (\mathbf{A}\,\xi_n, \xi_n) = \lambda_1 + \ldots + \lambda_n \,.$$

12.6 **Theorem:** Let $\lambda_1, \ldots, \lambda_n$ be the proper values of **A**, a linear transformation on E_n, and let $R\lambda_i$, $i = 1, \ldots, n$, be the real part of λ_i. Let $\mu_1 \geq \ldots \geq \mu_n$ be the real singular values of A. Then

(1) $R\lambda_i \leq \mu_1$, $i = 1, \ldots, n$,

(2) $R\lambda_i \geq \mu_n$, $i = 1, \ldots, n$, and

(3) $R\lambda_1 + \ldots + R\lambda_n = \mu_1 + \ldots + \mu_n$.

Proof: Let $\xi \in E_n$. Then we see that $(\mathbf{A}^*\xi,\xi) = (\xi,\mathbf{A}\xi) = \overline{(\mathbf{A}\xi,\xi)}$, where $\overline{(\mathbf{A}\xi,\xi)}$ is the complex conjugate of $(\mathbf{A}\xi,\xi)$. Now let ξ be a proper vector corresponding to the proper value λ_i of **A**, and let $|\xi| = 1$. Then

(4)
$$\left(\frac{\mathbf{A}+\mathbf{A}^*}{2}\,\xi, \xi\right) = \frac{1}{2}[(\mathbf{A}\,\xi, \xi) + (\mathbf{A}^*\,\xi, \xi)] = \frac{1}{2}[(\mathbf{A}\,\xi, \xi) + \overline{(\mathbf{A}\,\xi, \xi)}]$$

$$= \frac{1}{2}[\lambda_i(\xi, \xi) + \bar{\lambda}_i(\xi, \xi)] = \frac{1}{2}[\lambda_i|\xi|^2 + \bar{\lambda}_i\,|\xi|^2] = R\lambda_i \,.$$

On the other hand, by 12.2 (1) we have $R\lambda_i = \left(\frac{\mathbf{A}+\mathbf{A}^*}{2}\,\xi,\xi\right) \leq \mu_1$. This proves (1). Also, using 12.2 (2) with (4) we get $R\lambda_i = \left(\frac{\mathbf{A}+\mathbf{A}^*}{2}\,\xi,\xi\right) \geq \mu_n$.

To prove (3), we observe that $\mu_1 + \ldots + \mu_n$ is the trace of $\frac{\mathbf{A}+\mathbf{A}^*}{2}$. On the other hand, if $a_{11} + \ldots + a_{nn}$ is the trace of **A**, then $\bar{a}_{11} + \ldots + \bar{a}_{nn}$ is the trace of $\mathbf{A}^*$. It is easily seen, therefore, that the real part of the trace of **A** is equal to the trace of $\frac{\mathbf{A}+\mathbf{A}^*}{2}$, or

$$R\lambda_1 + \ldots + R\lambda_n = \mu_1 + \ldots + \mu_n \,.$$

12.7 **Theorem:** Let $\lambda_1, \ldots, \lambda_n$ be the proper values of **A**, a linear transformation on E_n, and let $\mathcal{I}\lambda_i$, $i = 1, \ldots, n$, be the imaginary part of λ_i. Let $\nu_1 \geq \ldots \geq \nu_n$ be the imaginary singular values of **A**. Then

(1) $\mathcal{I}\lambda_i \leq \nu_1$, $i = 1, \ldots, n$,

(2) $\mathcal{I}\lambda_i \geq \nu_n$, $i = 1, \ldots, n$, and

(3) $\mathcal{I}\lambda_1 + \ldots + \mathcal{I}\lambda_n = \nu_1 + \ldots + \nu_n$.

The proof is similar to the one in 12.6, and is left to the reader.

12.8 **Theorem:** Let **A** be a Hermitian transformation on E_n with proper values $\lambda_1, \ldots, \lambda_n$. Let $\{\xi_1, \ldots, \xi_n\}$ be any orthonormal set of vectors in E_n. Then

(1)
$$\begin{vmatrix} (\mathbf{A}\xi_1, \xi_1) & (\mathbf{A}\xi_1, \xi_2) & \ldots & (\mathbf{A}\xi_1, \xi_n) \\ (\mathbf{A}\xi_2, \xi_1) & (\mathbf{A}\xi_2, \xi_2) & \ldots & (\mathbf{A}\xi_2, \xi_n) \\ \ldots & \ldots & & \\ (\mathbf{A}\xi_n, \xi_1) & (\mathbf{A}\xi_n, \xi_2) & \ldots & (\mathbf{A}\xi_n, \xi_n) \end{vmatrix} = \lambda_1 \cdot \ldots \cdot \lambda_n \,.$$

Proof: Let the matrix of **A** with respect to $\{\xi_1, \ldots, \xi_n\}$ be

$$\begin{pmatrix} a_{11} & \ldots & a_{1n} \\ \ldots & & \ldots \\ a_{n1} & \ldots & a_{nn} \end{pmatrix}.$$

This means that $\mathbf{A}\,\xi_i = a_{i1}\,\xi_1 + \ldots + a_{in}\,\xi_n$, so that $(\mathbf{A}\,\xi_i, \xi_j) = a_{ij}$. This proves that (1) is

$$\begin{vmatrix} a_{11} & \ldots & a_{1n} \\ \ldots & & \ldots \\ a_{n1} & \ldots & a_{nn} \end{vmatrix} = \lambda_i \cdot \ldots \cdot \lambda_n .$$

12.9 **Theorem:** Let $\lambda_1, \ldots, \lambda_n$ be the proper values of **A**, a linear transformation on E_n. Let $\delta \geq \ldots \geq \delta_n$ be the absolute singular values of **A**. Then

(1) $|\lambda_i| \leq \delta_1$,

(2) $|\lambda_i| \geq \delta_n$, and

(3) $|\lambda_1| \cdot \ldots \cdot |\lambda_n| = \delta_1 \cdot \ldots \cdot \delta_n$.

Proof: Let $\xi \in E_n$ be a proper vector of **A** corresponding to λ_i such that $|\xi| = 1$. Then

$$(\mathbf{A}\,\xi, \mathbf{A}\,\xi) = (\lambda_i\,\xi, \lambda_i\,\xi) = \lambda_i\,\overline{\lambda}_i\,(\xi, \xi) = |\lambda_i|^2 .$$

But by 12.2

$$(\mathbf{A}\,\xi, \mathbf{A}\,\xi) = (\mathbf{AA}^*\,\xi, \xi) \leq \delta_1^{\,2} .$$

and also

$$(\mathbf{A}\,\xi, \mathbf{A}\,\xi) \geq \delta_n^{\,2} .$$

Therefore

$$|\lambda_i| \leq \delta_1 , \quad \text{and}$$

$$|\lambda_i| \geq \delta_n .$$

Now to prove (3) we observe that

$$\det(\mathbf{AA}^*) = \delta_1^{\,2} \cdot \ldots \cdot \delta_n^{\,2}$$

while

$$\det(\mathbf{A}) = \lambda_1 \cdot \ldots \cdot \lambda_n, \ \det(\mathbf{A}^*) = \overline{\lambda}_1 \cdot \ldots \cdot \overline{\lambda}_n ,$$

and since

$$\det(\mathbf{AA}^*) = \det(\mathbf{A}^*) \cdot \det(\mathbf{A})$$

the result is clear.

12.10 **Lemma:** For any Hermitian transformation **A**, the proper values of $\mathbf{A}^2 = \mathbf{AA}$ are respectively the squares of the proper values of **A**.

Proof: Since **A** is Hermitian $\mathbf{A}^* = \mathbf{A}$ and therefore $\mathbf{A}^2 = \mathbf{AA}^*$. Let ξ be a proper vector of **A** corresponding to the proper value λ.

Then

$$\mathbf{A}^2 \xi = \mathbf{A}(\mathbf{A}\xi) = \mathbf{A}(\lambda \xi) = \lambda \mathbf{A} \xi = \lambda^2 \xi .$$

Thus λ^2 is a proper value of $\mathbf{A}^2$.

Further, if ξ is a proper vector of $\mathbf{A}$ corresponding to λ, then ξ is also a proper vector of $\mathbf{A}^2$ corresponding to λ^2. Now let $\{\xi_1, \dots \xi_n\}$ be any orthogonal set of proper vectors of $\mathbf{A}$. Then $\{\xi_1, \dots, \xi_n\}$ is also an orthogonal set of proper vectors of $\mathbf{B} = \mathbf{A}^2$. If $\mathbf{B}\,\xi_i = \lambda_i\,\xi_i$, then $\mathbf{A}\,\xi_i = \mu_i\,\xi_i$, and $\lambda_i = \mu_i^2$.

12.11 **Theorem:** Let $\lambda_1, \dots, \lambda_n$ be the real singular values, $\mu_1, \dots, \mu_n$ be the imaginary singular values, and $\nu_1, \dots, \nu_n$ be the absolute singular values of $\mathbf{A}$, a linear transformation on E_n. Then

$$\lambda_1^2 + \dots + \lambda_n^2 + \mu_1^2 + \dots + \mu_n^2 = \nu_1^2 + \dots + \nu_n^2 .$$

Proof: It is observed that

$$\left(\frac{\mathbf{A} + \mathbf{A}^*}{2}\right)^2 + \left(\frac{\mathbf{A} - \mathbf{A}^*}{2i}\right)^2 = \frac{\mathbf{A}\mathbf{A}^* + \mathbf{A}^*\mathbf{A}}{2} .$$

By 12.10 we have

$$\text{trace} \left(\frac{\mathbf{A} + \mathbf{A}^*}{2}\right)^2 = \lambda_1^2 + \dots + \lambda_n^2 ,$$

$$\text{trace} \left(\frac{\mathbf{A} - \mathbf{A}^*}{2_i}\right)^2 = \mu_1^2 + \dots + \mu_n^2 .$$

By 12.4 we see that

$$\text{trace} \frac{\mathbf{A}\mathbf{A}^*}{2} = \text{trace} \frac{\mathbf{A}^*\mathbf{A}}{2} = \frac{\nu_1^2}{2} + \dots + \frac{\nu_n^2}{2}$$

It is easily seen that the trace of the sum of two matrices is the sum of the traces of those two matrices. I.e.,

$$\text{trace} \left(\frac{\mathbf{A} + \mathbf{A}^*}{2}\right)^2 + \left(\frac{\mathbf{A} - \mathbf{A}^*}{2i}\right)^2 = \lambda_1^2 + \dots + \lambda_n^2 + \mu_1^2 + \dots + \mu_n^2 .$$

$$= \text{trace} \frac{\mathbf{A}\mathbf{A}^* + \mathbf{A}^*\mathbf{A}}{2} = \nu_1^2 + \dots + \nu_n^2 ,$$

which proves the theorem.

There are many generalizations of the theorems given so far in this chapter. The purpose of this chapter however, was to give only a few samples of recent developments concerning singular values of a matrix and estimating proper values of a matrix from its singular values.

12.12 **The space of n-by-n matrices:** In 11.2 it was shown that the set of all n-by-n matrices with complex elements is a vector space. It is possible to define inner product and norm in this space in several ways. There are two norms which are in common use which are related to absolute singular values of matrices. In the following theorems we will give a few ideas about these relations.

12.13 **Hilbert norm:** Let $\mathbf{A}$ be a linear transformation on E_n. We define $||\mathbf{A}||_H$, the norm of $\mathbf{A}$, to be

$$\sup_{|\xi|=1} |\mathbf{A}\,\xi| .$$

By 12.12 we see that

$$||\mathbf{A}||_H^2 = \sup_{|\xi|=1} (\mathbf{A}\xi, \mathbf{A}\xi) = \lambda^2 ,$$

where λ is the largest absolute singular value of $\mathbf{A}$.

12.14 **Frobenius norm:** Let **A** and **B** be two linear transformations on E_n. Let

$$\begin{pmatrix} a_{11} & \dots & a_{1n} \\ \dots & & \dots \\ a_{n1} & \dots & a_{nn} \end{pmatrix}, \text{ and } \begin{pmatrix} b_{11} & \dots & b_{1n} \\ \dots & & \dots \\ b_{n1} & \dots & b_{nn} \end{pmatrix}$$

be the matrices of **A** and **B** with respect to an orthonormal base. We define (**A**,**B**), the inner product of **A** and **B**, as follows

$$(\mathbf{A},\mathbf{B}) = a_{11}\bar{b}_{11} + a_{12}\bar{b}_{12} + \dots + a_{nn}\bar{b}_{nn}, \text{ i.e.,}$$

the sum of the products of the form $a_{ij}\bar{b}_{ij}$ for $i, j = 1, \dots, n$. This clearly gives a norm for a matrix **A**, i.e.,

$$||\mathbf{A}||_F^2 = (\mathbf{A},\mathbf{A}) = |a_{11}|^2 + |a_{12}|^2 + \dots + |a_{nn}|^2 .$$

We shall prove that

$$||\mathbf{A}||_F^2 = \lambda^2 + \dots + \lambda_n^2 ,$$

where $\lambda_1, \dots, \lambda_n$ are the absolute singular values of **A**.

Let $\{\xi_1, \dots, \xi_n\}$ be the orthonormal base with regard to which the matrix of **A** is written. Then

$$\mathbf{A}\,\xi_1 = a_{11}\,\xi_1 + \dots + a_{1n}\,\xi_n ,$$

$$\dots$$

$$\mathbf{A}\,\xi_n = a_{n1}\,\xi_1 + \dots + a_{nn}\,\xi_n .$$

Therefore

$$|a_{11}|^2 + \dots + |a_{1n}|^2 = |\mathbf{A}\,\xi_1|^2 = (\mathbf{A}\,\xi_1, \mathbf{A}\,\xi_1) = (\mathbf{AA^*}\,\xi_1, \xi_1) ,$$

$$\dots$$

$$|a_{n1}|^2 + \dots + |a_{nn}|^2 = |\mathbf{A}\,\xi_n|^2 = (\mathbf{A}\,\xi_n, \mathbf{A}\,\xi_n) = (\mathbf{AA^*}\,\xi_n, \xi_n) .$$

Thus

$$|a_{11}|^2 + \dots + |a_{1n}|^2 + \dots + |a_{n1}|^2 + \dots + |a_{nn}|^2 = (\mathbf{AA^*}\,\xi_1, \xi_1) + \dots + (\mathbf{AA^*}\,\xi_n, \xi_n) .$$

But by 12.5, and 12.4

$$(\mathbf{AA^*}\,\xi_1, \xi_1) + \dots + (\mathbf{AA^*}\,\xi_n, \xi_n) = \lambda_1^2 + \dots + \lambda_n^2 . \text{ I.e.,}$$

$$||\mathbf{A}||_F^2 = \lambda_1^2 + \dots + \lambda_n^2 .$$

12.15 **Theorem** (*Proper values of the sum of Hermitian transformations*): Let **A** and **B** be two Hermitian transformations on E_n. Let $\lambda_1 \geq \dots \geq \lambda_n$ and $\mu_1 \geq \dots \geq \mu_n$ be respectively the proper values of **A** and **B**. Let $\nu_1 \geq \dots \geq \nu_n$ be the proper values of the Hermitian transformation **C** = **A** + **B**. [see 11.13] Then

$$\nu_{i+j-n} \geq \lambda_i + \mu_j \quad \text{where } i + j \geq n + 1 ,$$

or explicitly

$$\nu_n \geq \lambda_n + \mu_n ,$$

$$\nu_{n-1} \geq \begin{cases} \lambda_{n-1} + \mu_n \\ \lambda_n + \mu_{n-1} , \end{cases}$$

$$\ldots \qquad \ldots$$

$$\nu_1 \geq \begin{cases} \lambda_1 + \mu_n \\ \lambda_2 + \mu_{n-1} \\ \ldots \\ \lambda_n + \mu_1 \end{cases} .$$

Proof: By 12.2 (3) we see that for some subspace M of E_n, dim M = i, we have

$$\inf_{\substack{\eta \in M \\ |\eta| = 1}} (\mathbf{A}\eta, \eta) = \lambda_i .$$

Also for some subspace N of E_n with dim N = j we have

$$\inf_{\substack{\zeta \in N \\ |\zeta| = 1}} (\mathbf{B}\zeta, \zeta) = \mu_j .$$

But dim M + dim N = $i + j \geq n + 1$. This implies that there is a vector $\xi \in M \cap N$, $|\xi| = 1$, and for any such ξ,

(1) $$(\mathbf{A}\xi, \xi) \geq \lambda_i \text{ and } (\mathbf{B}\xi, \xi) \geq \mu_j .$$

Now by (10.15) we have

$$\dim (M \cap N) = \dim M + \dim N - \dim (M+N)$$

Thus dim $(M \cap N) \geq i + j - n$. This means that there is a subspace P of E_n such that $P \subset M \cap N$ and dim P = $i + j - n$. For some vector $\delta \in P$ with $|\delta| = 1$ we have

$$(\mathbf{C}\delta, \delta) = \inf_{\substack{\xi \in P \\ |\xi| = 1}} (\mathbf{C}\xi, \xi) \leq \nu_{i+j-n} .$$

But $\delta \in P \subset M \cap N$, and as in (1)

$$\nu_{i+j-n} \geq (\mathbf{C}\delta, \delta) = (\mathbf{A}\delta, \delta) + (\mathbf{B}\delta, \delta) \geq \lambda_i + \mu_j , \text{ or } \nu_{i+j-n} \geq \lambda_i + \mu_j .$$

12.16 **Theorem:** (*Wielandt*) Let **A** be any linear transformation on E_n. Let $\lambda_1, \ldots, \lambda_n$ be the absolute singular values of **A**. Let the matrix of **A** with respect to an orthonormal base be denoted by [**A**]. Then

$$[\mathbf{B}] = \begin{pmatrix} \mathbf{O} & [\mathbf{A}^*] \\ [\mathbf{A}] & \mathbf{O} \end{pmatrix}$$

is Hermitian and the proper values of **B** are precisely $\pm\lambda_1, \pm\lambda_2, \ldots, \pm\lambda_n$.

Proof: It is clear that [**B**] is Hermitian.

Let the absolute singular values of **A** be ordered as

$$\lambda_1 \geq \ldots \geq \lambda_k > \lambda_{k+1} = \lambda_{k+2} = \ldots = \lambda_n = 0 ,$$

and let $\{\xi_1, \ldots, \xi_k, \xi_{k+1}, \ldots, \xi_n\}$ be a complete orthonormal set of proper vectors of $\mathbf{AA}^*$ corresponding to $\lambda_1^2, \ldots, \lambda_k^2, \ldots, \lambda_n^2$. We define

$$\eta_i = \frac{\mathbf{A}\xi_i}{\lambda_i}, \; i = 1, \ldots, k .$$

Then

$$(\eta_i, \eta_j) = \frac{1}{\lambda_i \lambda_j}(\mathbf{A}\xi_i, \mathbf{A}\xi_j) = \frac{1}{\lambda_i \lambda_j}(\mathbf{AA}^* \xi_i, \xi_j) ,$$

and if $i \neq j$, then

$$(\eta_i, \eta_j) = \frac{\lambda_i^2}{\lambda_i \lambda_j}(\xi_i, \xi_j) = 0 ,$$

while

$$(\eta_i, \eta_i) = \frac{1}{\lambda_i^2}(\lambda_i^2 \xi_i, \xi_i) = 1 .$$

Therefore $\{\eta_1, \ldots, \eta_k\}$ is orthonormal. Completing this orthonormal set we get $\{\eta_1, \ldots, \eta_n\}$ which is a base for E_n. Now we prove

$$\mathbf{A}\xi_i = \lambda_i \eta_i \text{ and } \mathbf{A}^* \eta_i = \lambda_i \xi_i, \; i = 1, \ldots, n .$$

If $i = 1, \ldots, k$, then $\mathbf{A}\xi_i = \lambda_i \eta_i$ and $\mathbf{A}^* \eta_i = \lambda_i \xi_i$ are the direct result of the definition

$$\eta_i = \frac{\mathbf{A}\xi_i}{\lambda_i}.$$

If $i \geq k$, then $\mathbf{AA}^* \xi_i = 0$, and

$$0 = (\mathbf{AA}^* \xi_i, \xi_i) = (\mathbf{A}\xi_i, \mathbf{A}\xi_i) = |\mathbf{A}\xi_i|^2 ,$$

and this implies $\mathbf{A}\xi_i = 0$. On the other hand let ζ be orthogonal to $\eta_i, \ldots, \eta_k$. Then

$$0 = (\zeta, \eta_i) = \left(\zeta, \frac{\mathbf{A}\xi_i}{\lambda_i}\right), \; i = 1, \ldots, k .$$

Therefore $(\mathbf{A}^* \zeta, \xi_i) = 0$, $i = 1, \ldots, k$. Thus

$$\mathbf{A}^* \zeta = a_{k+1}\xi_{k+1} + \ldots + a_n \xi_n, \text{ and } \mathbf{A}^*\mathbf{A}\zeta = \mathbf{A}(\mathbf{A}^*\zeta) = a_{k+1}\mathbf{A}\xi_{k+1} + \ldots + a_n \mathbf{A}\xi_n .$$

Since $\mathbf{A}\xi_i = 0$, $i = k + 1, \ldots, n$ we have $\mathbf{A}^*\mathbf{A}\zeta = 0$. Hence

$$0 = (\mathbf{A}^*\mathbf{A}\zeta, \zeta) = (\mathbf{A}^*\zeta, \mathbf{A}^*\zeta) = |\mathbf{A}^*\zeta|^2 , \text{ i.e.}$$

$$\mathbf{A}^*\zeta = 0 \text{ which implies } \mathbf{A}^*\eta_i = 0, \; i = k + 1, \ldots, n$$

Now in terms of component, if $\xi_i = (x_{i1}, \ldots, x_{in})$ and $\eta_i = (y_{i1}, \ldots, y_{in})$, then let $\{\alpha_1, \ldots, \alpha_{2n}\}$ be the set defined as follows

$$\alpha_i = (x_{i1}, \ldots, x_{in}, y_{i1}, \ldots, y_{in}) \text{ for } i = 1, \ldots, n , \text{ and}$$

$$\alpha_i = (x_{i1}, \ldots, x_{in}, -y_{i1}, \ldots, -y_{in}) \text{ for } i = n + 1, \ldots, 2n .$$

Then $\{\alpha_1, \ldots, \alpha_{2n}\}$ is an orthogonal set. Let us denote by α_i for $i = 1, \ldots, n_1$ the matrix

$$\begin{pmatrix} x_{i1} \\ \cdot \\ \cdot \\ \cdot \\ x_{in} \\ y_{i1} \\ \cdot \\ \cdot \\ \cdot \\ y_{in} \end{pmatrix} = \begin{pmatrix} \xi_i \\ \\ \\ \eta_i \end{pmatrix}, \text{ and for } i = n+1, \ldots, 2n, \text{ the matrix } \begin{pmatrix} \xi_i \\ \\ \\ -\eta_i \end{pmatrix}.$$

Then we see that

$$\begin{pmatrix} \mathbf{O} & [\mathbf{A}^*] \\ [\mathbf{A}] & \mathbf{O} \end{pmatrix} \begin{pmatrix} \xi_i \\ \eta_i \end{pmatrix} = \begin{pmatrix} \mathbf{A}^*\eta_i \\ \mathbf{A}\,\xi_i \end{pmatrix} = \lambda_i \begin{pmatrix} \xi_i \\ \eta_i \end{pmatrix}, \text{ and}$$

$$\begin{pmatrix} \mathbf{O} & [\mathbf{A}^*] \\ [\mathbf{A}] & \mathbf{O} \end{pmatrix} \begin{pmatrix} \xi_i \\ -\eta_i \end{pmatrix} = \begin{pmatrix} -\mathbf{A}^*\eta_i \\ \mathbf{A}\,\xi_i \end{pmatrix} = -\lambda_i \begin{pmatrix} \xi_i \\ -\eta_i \end{pmatrix}.$$

This proves the theorem.

12.17 **Theorem:** Let $\lambda_1 \geq \ldots \geq \lambda_n$ and $\mu_1 \geq \ldots \geq \mu_n$ be respectively the absolute singular values of **A** and **B**, two linear transformations on E_n. Let $\nu_i \geq \ldots \geq \nu_n$ be the absolute singular values of $\mathbf{C} = \mathbf{A} + \mathbf{B}$. Then

$$\nu_{i+j-n} \geq \lambda_i + \mu_j \,, \quad i + j \geq n + 1$$

Proof: Combining the ideas of 12.15 and 12.16 we easily see the proof.

12.18 **Real or imaginary singular values of a sum of transformations:** Since

$$\frac{(\mathbf{A}+\mathbf{B}) + (\mathbf{A}+\mathbf{B})^*}{2} = \frac{\mathbf{A}+\mathbf{A}^*}{2} + \frac{\mathbf{B}+\mathbf{B}^*}{2}, \quad \text{and} \quad \frac{(\mathbf{A}+\mathbf{B}) - (\mathbf{A}+\mathbf{B})^*}{2i} = \frac{\mathbf{A}-\mathbf{A}^*}{2i} + \frac{\mathbf{B}-\mathbf{B}^*}{2i},$$

we easily see that the theorem 12.15 can be immediately applied to real and imaginary singular values of the sum of two transformations and a theorem similar to 12.17 can be stated and proved. It is left to the reader to state and prove the theorem.

EXERCISES 12

1. Find all the singular values of

$$\text{(i)} \begin{pmatrix} 1 & i \\ 0 & 3 \end{pmatrix}, \quad \text{(ii)} \begin{pmatrix} i & 2 \\ -1 & 1 \end{pmatrix}, \quad \text{(iii)} \begin{pmatrix} \frac{1+i}{2} & \frac{1-i}{2} \\ \frac{1-i}{2} & \frac{1+i}{2} \end{pmatrix}.$$

2. Let **A** be a Hermitian transformation. Show that the proper values of $\mathbf{A}^3 = \mathbf{AAA}$ are the cubes of the proper values of **A**.
3. Generalize problem 2 for $\mathbf{A}^n$, n a positive integer.
4.* Let $\lambda_1 \geq \ldots \geq \lambda_n$ be the proper values of **A**, a Hermitian transformation on E_n. Show that

$$\lambda_1 + \ldots + \lambda_k = \sup [(\mathbf{A}\,\xi_1, \xi_1) + \ldots + (\mathbf{A}\,\xi_k, \xi_k)] , \text{ and}$$

$$\lambda_n + \lambda_{n-1} + \ldots + \lambda_{n-k+1} = \inf [(\mathbf{A}\,\xi_1, \xi_1) + \ldots + (\mathbf{A}\,\xi_k, \xi_k)] ,$$

where $\{\xi_1, \ldots, \xi_k\}$ is any orthonormal set of vectors in E_n.
5. Let **A** be a positive transformation on E_n. Let $\lambda_1 \geq \ldots \geq \lambda_n$ be the proper value of **A**. Then

$$\lambda_1 \cdot \ldots \cdot \lambda_k = \sup \begin{vmatrix} (\mathbf{A}\xi_1, \xi_1) & (\mathbf{A}\xi_1, \xi_2) \ldots & (\mathbf{A}\xi_1, \xi_k) \\ \ldots & \ldots & \ldots \\ (\mathbf{A}\xi_k, \xi_1) & (\mathbf{A}\xi_k, \xi_2) \ldots & (\mathbf{A}\xi_k, \xi_k) \end{vmatrix} \quad \text{and}$$

$$\lambda_n \cdot \lambda_{n-1} \cdot \ldots \cdot \lambda_{n-k+1} = \inf \begin{vmatrix} (\mathbf{A}\xi_1, \xi_1) & (\mathbf{A}\xi_1, \xi_2) \ldots & (\mathbf{A}\xi_1, \xi_k) \\ \ldots & \ldots & \ldots \\ (\mathbf{A}\xi_k, \xi_1) & (\mathbf{A}\xi_k, \xi_2) \ldots & (\mathbf{A}\xi_k, \xi_k) \end{vmatrix} ,$$

where $\{\xi_1, \ldots, \xi_k\}$ is any orthonormal set in E_n.
6. Let **A** and **B** be linear transformations on E_n. Let $\lambda_1 \geq \ldots \geq \lambda_n$ and $\mu_1 \geq \ldots \geq \mu_n$ be respectively the absolute singular values of **A** and **B**. Let $\nu_1 \geq \ldots \geq \nu_n$ be the absolute singular values of $\mathbf{C} = \mathbf{AB}$. Prove that

(1) $\nu_{i+j-n} \geq \lambda_i\,\mu_j$, where $i + j \geq n + 1$

(2) $\nu_{i+j-1} \leq \lambda_i\,\mu_j$, where $i + j \leq n + 1$

7. Let **A** and **B** be Hermitian transformations on E_n with proper values $\lambda_1 \geq \ldots \geq \lambda_n$ and $\mu_1 \geq \ldots \geq \mu_n$ respectively. Let $\nu_1 \geq \ldots \geq \nu_n$ be the proper values of $\mathbf{C} = \mathbf{A} + \mathbf{B}$. Show that

$$\nu_{i+j-1} \leq \lambda_i + \mu_j,\ i + j \leq n + 1 .$$

APPENDIX

A review of solid geometry: We assume that the reader is familiar with the axioms of Euclidean geometry. What we are interested in here, is an intuitive discussion of lines and planes in the space (a three dimensional space). As the reader knows, a geometry is not a study of a single physical model of it. But we shall discuss the physical models of line and plane.

1. **The plane:** A plane is a surface having the following properties:

I. Through any two points of a plane passes a straight line whose points are all in the plane. Since a plane can be extended as much as desired we shall represent a plane by a portion of it which is bounded by a rectangle. Since a right angle in perspective is not seen as a right angle we shall draw a parallelogram for a plane (Fig. I).

Fig. I

II. Three points which are not on the same line determine a unique plane.
III. Two parallel lines determine a unique plane.
IV. Two intersecting lines determine a unique plane.
V. A line and a point which is not on the line determine a plane.

Some of these properties are axioms and others are theorems. We shall not go into that. We expect the reader to consult a formal book in solid geometry.

2. **Comparison of a line and a plane:** We have already discussed the fact that two points of a line being in a plane implies that the line lies entirely in the plane.

I. If a line has only one point A in common with a plane $\mathscr{P}$ we say it intersects $\mathscr{P}$ at A (Fig. II).
II. If a line does not intersect a plane, we say the line is parallel to the plane (Fig. III).
III. If a line l is parallel to a line d in the plane $\mathscr{P}$, then l is either in the plane $\mathscr{P}$ or parallel to it (Fig. III).
IV. If a line l is perpendicular to two lines of a plane $\mathscr{P}$ at the point A of intersection of the lines, then any line of the plane through A is perpendicular to l. This is a very important theorem. We shall give a figure for it (Fig. IV). For the proof see any standard book in solid geometry.

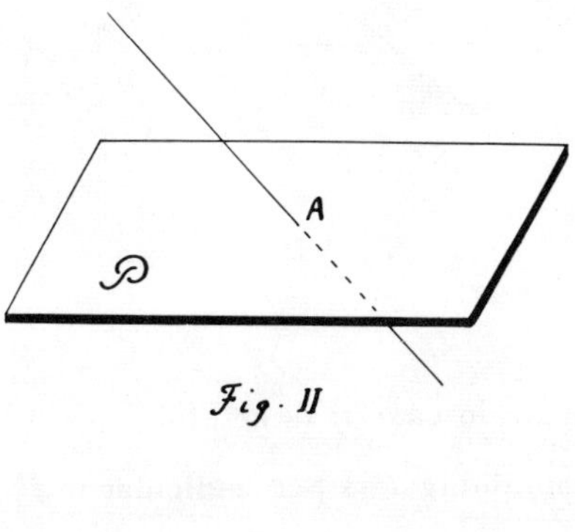

Fig. II

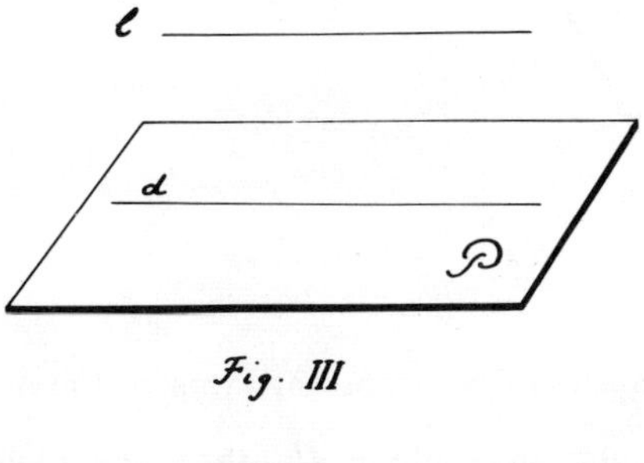

Fig. III

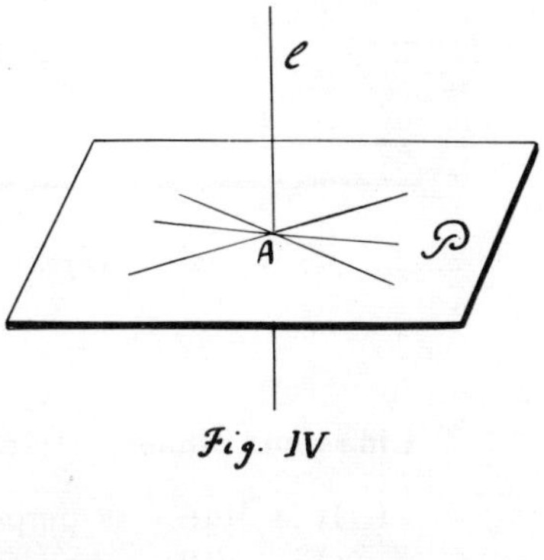

Fig. IV

The line l is defined to be perpendicular to $\mathcal{P}$, and $\mathcal{P}$ is called a plane perpendicular to l. We see easily that through a point P there is one and only one line perpendicular to a plane $\mathcal{P}$. Also through a point **P** there is a unique plane perpendicular to a given line l. (What has been discussed here is used in 1.19 to obtain the equation of a plane.)

3. **Two planes:** First we shall state a very important axiom: *If two planes have a point* **A** *in common, they have a line in common* (Fig. V). The two planes may be coincident.

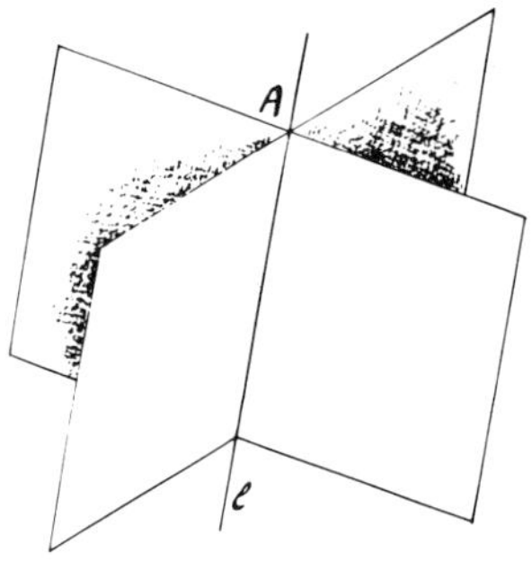

Fig. V

I. Two planes are called parallel if they do not intersect. The reader may show that through a given point there is one and only one plane parallel to a given plane (Fig. VI).

II. The angle between two planes: Let $\mathcal{P}_1$ and $\mathcal{P}_2$ be two planes which intersect in a line l (Fig. VII). Choose a point **A** on l. Construct a perpendicular to l at **A** in $\mathcal{P}_1$ and a perpendicular to l at **A** in $\mathcal{P}_2$. The angle between these two lines is defined to be the angle between $\mathcal{P}_1$ and $\mathcal{P}_2$. This definition needs justification. We should show that the measure of this angle is independent of the choice of **A**.

III. If the angle between two planes is a right angle, we say the two planes are perpendicular.

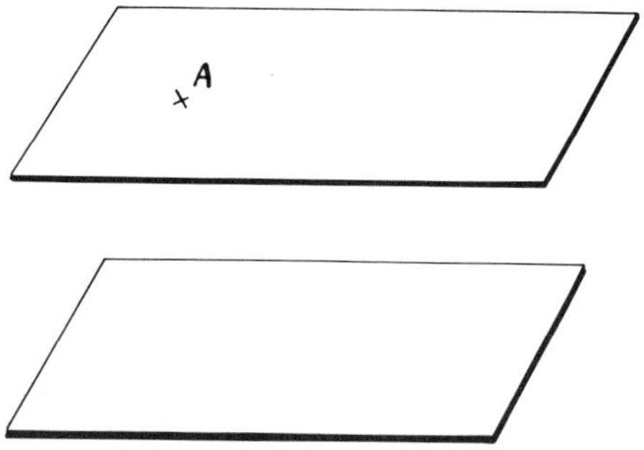

Fig. VI

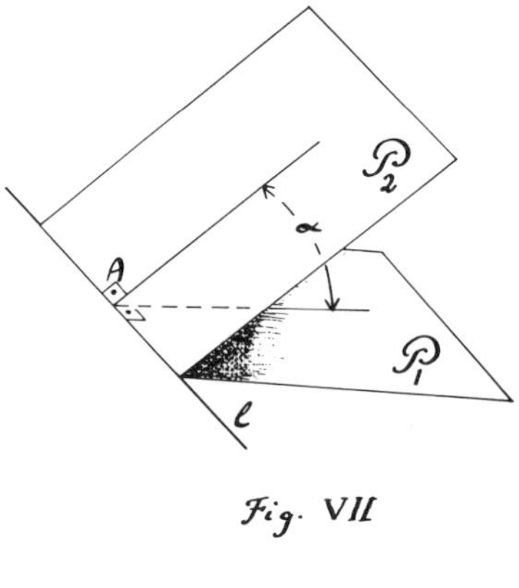

Fig. VII

4. **Lines and planes:** Here we state a few important facts and give a figure in case it is helpful.

I. If a line l is perpendicular to a plane $\mathcal{P}$, then any plane containing l is perpendicular to $\mathcal{P}$ (Fig. VIII).

II. Let $\mathcal{P}_1$ and $\mathcal{P}_2$ be two parallel planes. Then any plane which intersects $\mathcal{P}_1$ will intersect $\mathcal{P}_2$ and the lines of intersection are parallel (Fig. IX).

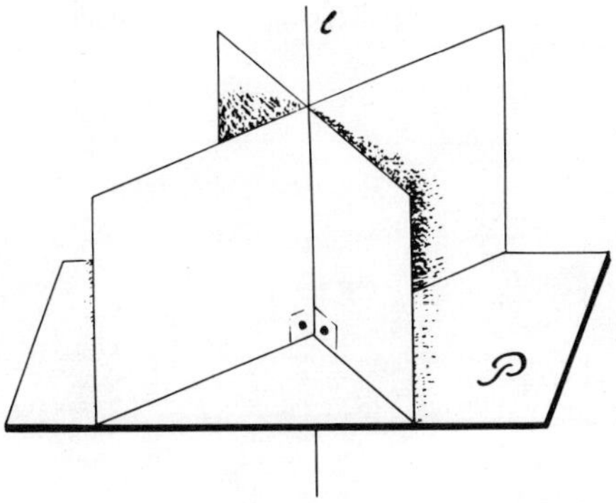

Fig VIII

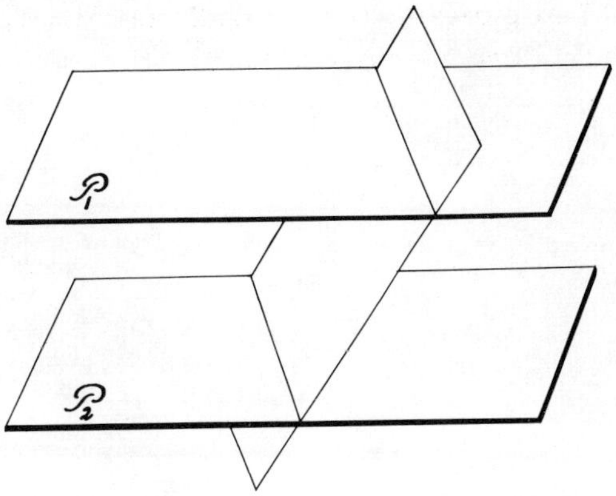

Fig. IX

5. **Skew lines:** Two lines which do not intersect and are not parallel are called skew.

Problem: Construct the common perpendicular of two skew lines l and m.

Solution: Through a point **A** of m we draw a line parallel to l (Fig. X). This line with m gives the plane $\mathcal{P}_1$ which is parallel to l. Through a point **B** of l draw a line parallel to m. This line with l gives a plane $\mathcal{P}_2$ which is parallel to m and $\mathcal{P}_1$. Construct a plane Q through l perpendicular to $\mathcal{P}_1$. This is done by choosing a point **B** on l and drawing a line through **B** perpendicular to $\mathcal{P}_1$. This line with l will give Q.

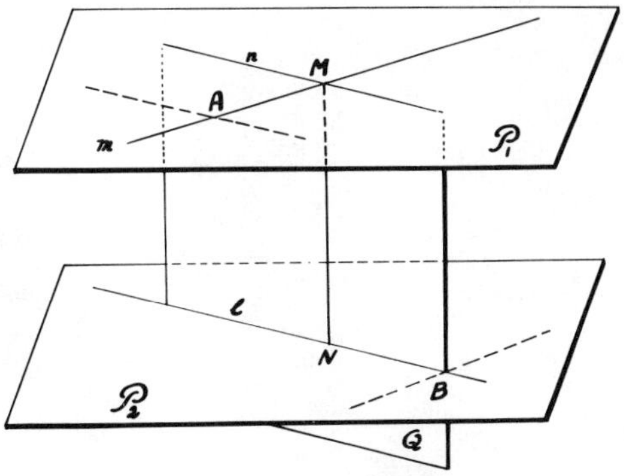

Fig. X

The plane Q intersects $\mathcal{P}_1$ in a line n parallel to l. If **M** is the point of intersection of m and n, then the line through **M** perpendicular to $\mathcal{P}_1$ is the desired line. We leave it to the reader to complete the proof. Analytic solution of this problem is asked in Exercises 9 No. 2.

6. **Projection onto a plane:** Let $\mathcal{P}$ be a plane and **P** a point. The projection of **P** on $\mathcal{P}$ is **Q**, the foot of the perpendicular through **P** to $\mathcal{P}$.

The projection of a cofiguration on $\mathcal{P}$ is the set of the projections of all its points.

INDEX